Breaking Bed Bugs: How to Get Rid of Bed Bugs without Losing Your Mind, Money & Dignity

© 2016 Chipp Marshal

All rights reserved. No part of this publication may be reproduced, distributed, or transmitted in any form or by any means, including photocopying, recording, or other electronic or mechanical methods, without the prior written permission of the author and publisher.

The contents of this book are based on personal experience and extensive research. The author and publisher have made every effort to ensure that the information in this book was correct at the time of publication. However, the author and publisher do not assume and hereby disclaim any liability to any party for any loss, damage, or disruption caused by usage of products, errors or omissions, whether such errors or omissions result from negligence, accident, or any other cause. While the author is well educated and experienced on this topic, he does not claim to be a licensed professional. Apply this information at your own risk and please use common sense when utilizing any of the information contained in this publication.

Table of Contents

Foreword 3

Introduction 4

My Bed Bug Story 6

Chapter 1: Bed Bug Information and History 11

Chapter 2: The Psychological and Emotional Effects of Bed Bugs 14

Chapter 3: Should I Really Do This Myself? 16

Chapter 4: Verifying a Bed Bug Infestation (The Initial Inspection) 17

Chapter 5: Determining the Extent of the Infestation (The Full Inspection) 22

Chapter 6: Treatment Preparation 27

Chapter 7: The Importance of Containment 31

Chapter 9: Should You "Protect" or "Isolate" Your Bed? 37

Chapter 10: Treating the Bed 40

Chapter 11: Vacuuming Instructions 48

Chapter 12: Steam Instructions 52

Chapter 13: 100% Amorphous Silica Gel Desiccant Dust (ASG Dust) 55

Chapter 14: Organization, Containment, and Treatment of Belongings ... 57

Chapter 15: It's Time to Kill Some Bed Bugs! 61

Chapter 16: Pest-Proofing Rooms 65

Chapter 17: Monitor and Maintain 68

Chapter 18: Conclusion 70

Frequently Asked Questions 71

Bed Bug Supply Manual 78

Foreword

I was delighted to hear that Mr. Marshal was publishing his methods for getting rid of bed bugs. As a real estate agent and property manager, I see bed bug infestations more frequently with each passing year. Many of the affected families cannot afford to hire an exterminator, so parents feel helpless and children suffer. For some, there is no escape from this epidemic and it is absolutely heartbreaking to watch.

This book is being published by the right person at the right time. It is reassuring to know that his methods will now be available to millions of people around the world.

—*Christina Murray*

Introduction

"Good night. Sleep tight. Don't let the bed bugs bite!"

Besides this catchy little nighttime jingle, hardly anyone wants to talk about bed bugs. "If I let someone know I have bed bugs," we reason, "then they will think that I am dirty."

The truth is that bed bugs are not caused by poor sanitation, and they have nothing whatsoever to do with social or economic status. They are cunning hitchhikers, and *anyone* who is in the wrong place at the wrong time is susceptible to an infestation. They can thrive in the lowliest hovel as well as in the finest five-star hotels, well run hospitals, and million-dollar homes.

If you're reading this book, you probably know—or suspect—you have a bed bug problem. If that's the case, don't lose heart. There is hope.

I know firsthand the devastation bed bugs can have on your physical, emotional, and mental well-being. You can read my personal bed bug story in the next section of this book.

After I won the battle in my own home, I used my knowledge and experience to successfully guide hundreds of other families in getting rid of bed bugs in their homes. I am here to tell you that you, too, can get rid of bed bugs yourself—quickly, safely, and inexpensively.

I will share with you in this book everything I wished someone had told me about getting rid of bed bugs. I will tell you all those secrets an exterminator won't share with you. I'll give you the details the websites and videos you'll find on the internet leave out. I will do everything in my power to prepare you mentally, physically, and emotionally for the task at hand.

You stand at a turning point. Half measures will avail you zero results. The methods I'll describe for you have been tested and proven. To ensure they work you will need to take action, following the instructions precisely. If you decide to fudge on one detail or another, or skip a step or two, you will most likely end up right back where you started.

With this book, I give you a proven course of action. You are required to supply the courage and resolve. Gird your loins for battle.

My Bed Bug Story

My bed bug story began in the Fall of 2010. The summer was coming to an end and my firstborn son had just started kindergarten. It was an exciting time for our family. I'll always remember seeing my son get on the bus for his first day of school. Unfortunately, my family wouldn't be able to enjoy that happy time for very long.

I woke up one morning with a few itchy bumps on my left leg. It didn't worry me much so I went about my day as usual and didn't give it anymore thought.

The next morning, I had more of the same itchy marks. They were on both of my legs this time. I proceeded to do what everyone does in this day and age and tried to self-diagnose my symptoms on the internet!

It could have been anything at that point. I did come across "bed bugs" as a possible cause, but I immediately dismissed that idea. There was no way we could have had bed bugs, I thought. I ended up convincing myself that it was a mild allergic reaction or rash.

The marks continued to appear in different areas over the next few days. I found myself back on the internet after trying every cream and ointment in the medicine cabinet. Again, "bed bugs" kept popping up as a possible cause. I didn't think it was a possibility but I decided to do an inspection just to rule it out. Looking back, it wasn't a thorough inspection because I didn't know anything about bed bugs at the time, but it was enough for me to cross them off my list.

At this point I was having trouble sleeping through the night. The itching was driving me crazy. I scheduled a doctor's visit but the doctor couldn't see me for another few days.

Then, in the middle of one night my five-year-old son walked into my bedroom and announced, "Daddy, I am itchy." I picked him up and turned the lights on in my room. There they were. I saw a few little bugs crawling on my bed sheet. My heart sank because I immediately knew what they were.

We had bed bugs.

It was a helpless feeling. My wife and our two-year-old son were both still sleeping. I was holding my son, standing there in the middle of the night wondering what to do. I felt horrible that my children were not safe in their own home. My five-year-old had more than twenty bites on his arms. I felt like a terrible father. Could I have prevented this? How do I fix it? What should I do right now?

I woke my wife up and she checked on our younger son. He was sound asleep, but still the thought of him being bitten by disgusting bugs was killing me inside. Neither my wife nor our younger son was showing symptoms of being bitten, but as we learned later, not everyone shows a reaction to bed bug bites.

This was one of the most helpless feelings a father could ever feel. My family was suffering and I couldn't do anything to help them. I knew from the little bit of reading I did about bed bugs that I shouldn't sleep somewhere else in the house or leave for a hotel.

I wasn't putting my kids back in their beds. Hell, I wasn't getting back in my bed either.

I finally decided to grab our sleeping bags and camp out on the trampoline in the backyard for the night. No one said we couldn't do that! Besides, we camped out on the trampoline a few times over the summer so my kids were at least familiar with the idea.

As a summer activity it was exciting to sleep under the stars. As a temporary refuge from parasitic insects, it was heartbreaking. I didn't sleep a minute for the rest of the night.

The following morning I didn't waste any time, and I called an exterminator immediately. The first company said they couldn't send anyone to my home for two days. The next company I called said the same thing. The third one couldn't come out until the following week. Surely these people understand what we are going through, I thought. I needed an emergency response team at my home immediately!

I was finally able to get hold of someone who could come to my home that same day. When he got there, he found a large number of bed bugs in our box spring. There was also evidence of bed bugs in both of my sons' beds.

He eventually hit me with the price for treatment. Wow. At that point, I didn't think I had any choice but to find the money and pay his fee.

For six weeks we followed all of his instructions. We threw stuff away, washed clothes, and bagged almost everything we owned.

The technician did three treatments, each two weeks apart. He showed up, sprayed chemicals all over the place, steamed the bed, and left. The only thing we had to guide us during the rest of the time was a little pamphlet he gave us about doing our part to make sure the treatment was successful.

At the end of those six weeks, he told us the bed bugs were gone. It was finally over!

Or was it?

Two weeks later I saw a bed bug scurry across the bed. My heart sank again.

I called the exterminator to inform him that the bed bugs were back. I'll never forget his priceless response. He offered us a "discount" off his regular price since we were "repeat" customers. He went on to explain that he held up his end of the bargain and if we still had bed bugs it was either a new infestation or we did something wrong during the treatments.

What I said to him next does not count as one of the prouder moments in my life.

Fortunately, I had observed everything he did during the course of his treatments. I always asked for details about the products and strategies he used. I had also been reading every piece of research I could find about how to deal with bed bugs.

I couldn't afford to hire someone else at this point. I knew I probably had legal recourse, but how long would that take to resolve? I needed to do something for my family right then.

No one else cared more about getting bed bugs out of my home than I did. So I decided that made me the best candidate for the job.

I would have my work cut out for me, though. It was a very tough time to start treatments on my own because everything the exterminator had sprayed up until that point had likely driven the bed bugs deeper into our home.

I took a few of the techniques I learned from the exterminator and supplemented them with better strategies I discovered in my research. I scoured forums with bed bug-related discussions. I also reached out to an entomologist for advice.

There was a great deal of trial and error in the beginning as I refined my procedures. In fact, as I write this guide, I get a good laugh at the way I originally did certain things compared to the techniques I teach now.

Nonetheless, with hard work, faith, and persistence, our bed bug nightmare finally came to an end.

We would eventually transition back to normal life. However, there was definitely a degree of post-traumatic stress and lots of paranoia for the next few months.

It would take a few weeks before we stopped sleeping with the lights on. It took me about 3 months before I stopped thinking about bed bugs every waking moment of the day. Time would eventually heal all of our emotional wounds. We were able to feel safe and comfortable in our home again. We even reached a point where we could laugh about some of our hardships. We were happy again.

I kept this story to myself for a long time. Writing this book is the first time I have shared it outside of my family and close friends. If you have discovered you have bed bugs, too, I want you to know that I understand how you feel and what you are going through. Hang in there and have faith that you will get through it. If we could do it, so can you.

Chapter 1: Bed Bug Information and History

"Why do I need to know about the history of bed bugs?" you ask.

I would love nothing more to give you the first step of the treatment process right now. However, you will need some basic information about bed bugs to form a foundation of knowledge for the steps that lie ahead. Think of it as a sporting event. You need to develop a specific mindset in order to successfully defeat your opponent. In this chapter I'll give you the low-down dirt on bed bugs so you can understand what you are up against and what it is going to take to win.

Bed bugs are not some recent phenomenon hyped up by the media. They have been around for thousands of years. In fact, there is evidence that the Romans and ancient Egyptians were plagued by bed bugs.

Bed bug infestations are not new to America, either. They seem to have hitched rides with the colonists in the 17th century on their voyages across the Atlantic Ocean.

Bed bugs thrived in America until the 1950's when they were nearly eradicated by the use of extremely toxic pesticides. However, those chemicals pose serious threats to human and environmental health, and new regulations on those pesticides have severely limited their use in recent years.

Once the toxic chemicals were restricted, bed bugs returned with a vengeance and are thriving in the modern world. Besides necessary restrictions on the use of dangerous insecticides, there are numerous other factors that have contributed to the resurgence of bed bugs.

- Bed bugs are now highly resistant to most modern pesticides and insecticides.
- Bed bugs are skillful hitchhikers who hide in luggage and clothing, giving them free access to every part of the globe now that we have worldwide travel.

- Bed bugs reproduce quickly and can build large populations that go undetected for critical lengths of time.
- Most people lack awareness about bed bugs, as they know little or nothing about these pests.
- Until very recently, most people—including professional pest control operators—had no experience in dealing with bed bugs.
- Until now, we haven't taken the problem seriously as a society.

Will We Win Against Bed Bugs?

Bed bugs have finally become enough of a problem in the United States that they have Americans' attention, however, and it has become apparent to nearly everyone that a viable large-scale solution to the problem is necessary. Various levels of government, scientists, pest control companies, and the hospitality and travel industries are starting to come together to find answers.

Still, most people lack the knowledge and resources to deal with the problem themselves. To make matters worse, the travel and hotel industries appear to place a disproportionate emphasis on response strategies rather than preventative ones. This worries me.

I believe we will start winning the war on bed bugs only when our focus shifts to preventing bed bugs from establishing a foothold in the first place. Until then, we are forced to take defensive actions in order to fend them off. Unfortunately, then, it is my opinion that the bed bug situation in the United States will continue to get worse for the foreseeable future.

Chapter 2: The Psychological and Emotional Effects of Bed Bugs

Bed bugs are well known for their bites. However, they inflict the most damage through psychological and emotional wounds. Is there anything worse than being assaulted in the place where we expect to feel the safest?

After bed bugs move into your home, your nights are sleepless ones, permeated with anxiety and paranoia. Night after night, the stress compounds, with a cumulative effect similar to—and as severe as—post-traumatic stress disorder. If you allow them to, these effects can linger long after the bed bugs are gone.

Keep Your Head Up

I can't stress enough that *everyone* who leaves their home is prone to a bed bug infestation. Social status and cleanliness play no roles whatsoever. So keep your head up and tackle this problem head on. There is no reason to be ashamed that bed bugs chose your home. They're not particular. They'll go anywhere, whether they're welcome or not.

Maintain Control of Your Life

The following steps will help you to maintain control of your life while you're dealing with a bed bug infestation:

- Keep things in perspective. Dealing with a bed bug infestation can be an overwhelming process. Remind yourself frequently that this is only a temporary problem, a phase in your life.

- Be diligent in your removal efforts and follow the steps in this book carefully. If you can see your progress your outlook on the situation is bound to improve.

- Do not isolate yourself or your family.

- Utilize your support network. Talking about the problem with people you trust will help you to deal with the anxiety and stress.
- It is usually best to be honest with friends and family. This will allow them to protect themselves and prevent an infestation from spreading to their homes, too.
- Try to get out and enjoy life. It is important to find activities that will give your mind time to relax.
- Never be afraid to ask for help.

Do Not Allow Fear to Control You

Sometimes physically getting rid of bed bugs is only half of the battle. If you are struggling with psychological and mental issues that affect your everyday life, seek professional counseling. I have heard from a number of people that speaking to a professional really helped them. If you don't have insurance, your family doctor can direct you to a resource that will provide services for no fee or fees on a sliding scale. I fully understand the stress and hardship bed bugs can cause. However, I promise you that you can and will enjoy life again.

In severe cases, some people report experiencing major depression and even suicidal thoughts. If you are feeling suicidal or thinking of harming yourself, please go to an emergency room or contact a professional immediately. Always remember that it OK to ask for help. You can also contact the National Suicide Prevention Lifeline at 1-800-273-TALK.

Chapter 3: Should I Really Do This Myself?

There are two quotes about "half measures" that I have grown fond of in my life, and I apply their philosophy to every important goal I set out to accomplish. The first of those quotes is from the Alcoholics Anonymous *Big Book*:

"Half measures availed us nothing. We stood at the turning point."

In critical situations, half measures avail us zero results, and this is especially so when it comes to treating bed bugs. One hundred percent of your commitment and resolve will be required to get you past the turning point where you stand right now.

The second quote is from my one of my favorite television shows, *Breaking Bad*. The fictional character Mike Ehrmantraut, head of corporate security at Los Pollos Hermanos, is speaking to Walter White about his past mistakes:

"I chose a half measure when I should have gone all the way. I'll never make that mistake again. No more half measures, Walter."

(Jonathan Banks and Bryan Cranston are the actors who portrayed the two characters in that series.)

I don't' want to waste your time, so why am I talking about quotes and movies? That's because I believe it is extremely important that you understand—and adopt—the mindset required for accomplishing the task that lies before you.

So long as you are committed to using the full measures as described in this book, you are fully capable of getting rid of bed bugs by yourself.

Chapter 4: Verifying a Bed Bug Infestation (The Initial Inspection)

About a third of people who are bitten by bed bugs show no physical signs of the bites. Likewise, there are a number of physical reactions to other insects, bacteria, and allergens that could be mistaken for bed bug bites.

Therefore, never use the presence of bite marks—or the lack of them—to determine whether or not you have a bed bug infestation.

Excrement, Blood Spots, and Eggs

The most accurate way to verify an infestation is by detecting physical signs of the bugs themselves. Use a magnifying glass and an ultraviolet (UV) light source to inspect for bed bugs' excrement, blood spots, and eggs:

- If you have an infestation, you will likely discover excrement, blood spots, and eggs.

- Bed bug feces will look like a tiny dark spot about the size of the period in this sentence.

- Blood spots will be small blood stains from crushed bed bugs on your bedding or mattress.

- Eggs will appear white or transparent in color about the same size as the excrement. (2-3mm in length)

Image by flickr user Louento.Pix (CC BY-ND 2.0)

While these signs will help in the verification process, none of them are one hundred percent reliable in identifying a bed bug infestation. Each of these items could be something else entirely, and this evidence taken alone may lead you to falsely believe that you have a bed bug presence.

Skin Castings and Live Bed Bugs

In order to have complete certainty, verify a bed bug infestation by locating a live bed bug or skin castings. Bed bugs cast their skins as they grow. These castings vary a great deal in size, but you can learn to recognize them. They look like a paper version of a live bed bug.

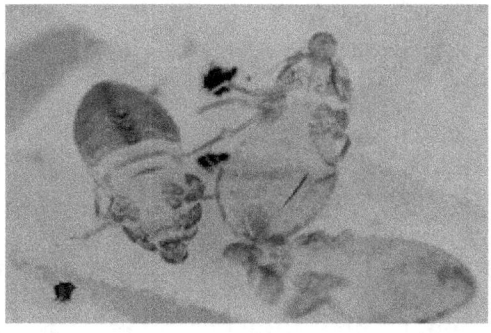

Image by flickr user Louento.Pix (CC BY-ND 2.0)

What Do Bed Bugs Look Like?

Adults:

- Adult bed bugs are small, oval-shaped insects, complete with six legs and antennae.

- Before feeding, they measure roughly 5-6mm (about 1/4 inch) long. They appear flat in shape and reddish-brown in color.

- After a meal, they expand to 7-10mm (about 3/8 inch) long and turn bright red in color.

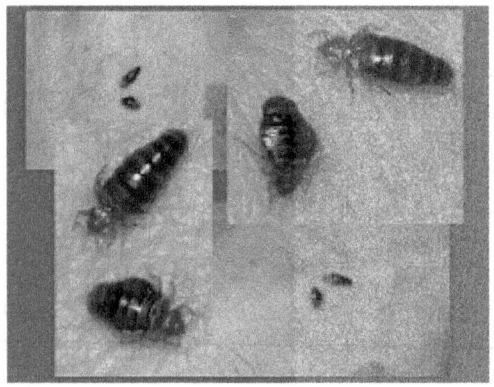

Image by flickr user Louento.Pix (CC BY-ND 2.0)

Youth:

- Nymphs (juveniles) range in size from 1mm to 4mm (barely visible to about 1/8 inch).

- Prior to feeding they are a whitish/transparent color.

- After feeding, nymphs turn a bright red blood color.

Identification Tips

It is important to observe as many photos as possible to familiarize yourself with the appearance of bed bugs. Use the resources in this guide along with additional online photo searches. You can also utilize other online resources such as:

www.reddit.com/r/bedbugs or **www.reddit.com/r/whatsthisbug**

When observing bed bugs yourself or taking a photo to post on Reddit.com, it is best to place the bed bug on a white surface to make identification of its features easier. A photo from the top, bottom, and side, taken with adequate lighting, is also helpful if you are posting to reddit.com for confirmation.

Tools such as a flashlight and magnifying glass will make identification easier.

Other Bugs and Items That Are Commonly Mistaken for Bed Bugs

Making bed bug identification even trickier is the fact that so many other bugs are similar in appearance, as are tiny pieces of debris that almost everyone has in their home. Below is a list of common items and other bugs that can be confused with bed bugs, their nymphs, and their eggs:

- Bat bugs
- Spider beetle
- Tick bugs
- Larder beetle larva
- Juvenile German cockroaches
- Dried skin
- Dried nasal mucus
- Pebbles, seeds, and other tiny pieces of debris

Where to Look

Bed bugs will routinely travel up to twenty feet from their established harborages to feed. Most infestations can be found around the bed. Please remember that bed bugs can easily squeeze into really small hiding spots. If you can slip the edge of a credit card into a crack or crevice, it is large enough to harbor bed bugs.

Focus your search on the seams and crevices of the mattress and box spring. Other areas you should inspect include (but are not limited to):

- Seams and tags of mattresses and box springs
- Cracks, crevices, and screw holes on the bedframe, headboard, and walls surrounding the bed
- Seams and folds of chairs, couches, cushions, and curtains
- Furniture joints such as on a dresser or night stand
- Inside appliances
- Picture frames
- Electrical receptacles
- Carpet seams where the floor meets the wall
- Drywall seams where walls and ceilings meet

Chapter 5: Determining the Extent of the Infestation (The Full Inspection)

One of the most common reasons for control failures when it comes to bed bugs is failure to fully inspect and find all the bugs' hiding places. The purpose of the full inspection is to gather critical information about the locations and extent of the infestation. This information will be very helpful with subsequent treatment measures.

Start by noticing where in your home you are most often bitten and where you have already seen bed bugs. Take into account your movements and sleeping patterns. If you have slept in various areas within your home, then all of these areas need to be inspected and treated.

The University of Kentucky conducted a study to determine where you are most likely to find bed bugs in your home. Use these percentages for informational purposes only. Every room and every infestation is different.

These results are based on a study of 13 apartments infested with bed bugs. This is where the bed bugs were found:

- 34.6% - box spring
- 22.6% - furniture such as couches and upholstered chairs
- 22.4% - mattress
- 13.4% - bed frame and headboard
- 3.1% - baseboard areas
- 2.3% - walls and ceilings
- 1.4% - other
- .2% - nightstands and dressers

The Inspection Process

Bed bugs have a very flat body shape which allows them to hide in virtually any crack or crevice. They prefer dark, isolated, and protected areas, and they like wood, paper, and fabric surfaces, so pay special attention to these areas and these materials.

Inspect your mattresses first and pay extra close attention to the following areas:

- Seams
- Labels
- Corner protectors
- Beading
- Buttons

Carefully inspect adjacent components such as:

- Hollow plastic caster legs
- Bed spring coils
- Interiors of hollow bed posts
- Bedframe
- Headboard
- If the bed has wooden slats, check each slat carefully, as wooden slats contain many cracks for bed bugs to hide in and lay their eggs.
- If the wooden slats are bolted to the bedframe, the bolts should be undone and the drilled holes inspected and treated.
- If your bed is an ensemble bed, check the base carefully as it is more likely to harbor the bugs.
- Be sure to check the edge of the material underneath the ensemble base.

The areas around the bed should be investigated next:

- Bedside furniture such as tables, dressers, or other furniture should be turned over and examined.
- Drawers in bedside furniture should be removed and examined.
- If headboards are attached to the wall, they should be removed to allow access for inspection and treatment.

Once you have thoroughly inspected the bed, move on to inspect the rest of the bedroom, checking other items and furniture throughout the room and paying close attention to seams, buttons, and wooden areas. Here is a checklist of some easily overlooked items and areas:

- Luggage
- Locations where luggage is kept
- Appliances such as telephones, alarm clocks, radios, etc.
- Books
- Clothes and laundry (both clean and dirty)
- Behind switch plates and electrical outlets
- Smoke detectors
- Light fittings
- Underneath carpet edges
- Underneath door transitions
- Underneath rugs
- Skirting boards, joins in floor boards
- Loose wall paper and paint
- Old nail and screw holes in walls and ceilings
- Cracks in walls and ceiling

- Any other wall voids
- Moldings
- Window casings
- Picture frames, mirrors, and any other wall hangings

Keep records that indicate all locations of bed bug activity. Most bed bug control instructions will tell you to inspect adjoining rooms next to, above, or below the areas of a detected infestation. My advice is to inspect your entire house. It just isn't worth using a half measure when you should go all the way. If all harborages are detected, control and treatment will be easier and more successful.

High-Risk Factors That Could Affect Your Treatment Efforts

Certain risk factors at the site of the infestation can make the job extremely challenging and increase the likelihood of treatment failure. These include:

- **Clutter**: Bed bug control in cluttered rooms is nearly impossible. If you have a lot of clutter, I will guide you through where to put it in a future chapter.
- **Cooperation**: It will be your job to ensure that other members of the home understand that successful bed bug treatment will require their cooperation. This also includes cooperation from a landlord if you are a renter as well as cooperation from neighbors if you live in a multiple occupancy dwelling.
- **Construction and Room Integrity**: Certain building materials such as exposed brick provide more challenging treatment conditions. Likewise, excessive gaps, cracks, and crevices in and around walls, ceilings, floors, trim, and molding can all make bed bug extermination more difficult.

- **Multi-Unit Dwellings**: Treating apartments, duplexes, and condominiums is often more challenging due the higher likelihood of a re-infestation.

- **Heavy Infestations**: Heavy infestations will naturally require more treatment and time.

- **Improper Control Attempts**: If you have already attempted to get rid of your bed bug infestation by using improper control techniques, there is a good chance that you made things worse by spreading the infestation deeper into your home. Some exterminators will not even take the job if there is evidence that you used improper treatment techniques such as a "bug bomb." Another way you can make things worse is by trying to get rid of bed bugs with your vacuum cleaner without first setting up proper room containments and following the other instructions in this book. Take things step by step and you'll see better results.

Chapter 6: Treatment Preparation

I encourage you to gather all of the supplies you will need before you begin treating the infestation. It's best to have everything on hand so you can quickly and seamlessly move from one step to the next. Doing so will play a critical role in the success rate of these treatments.

Remember, this process will take some patience and may require you to repeat certain steps. No single product or method exists that will eliminate bed bugs with one swift action. Only an integrated approach with multiple methods of treatment can be successful. Don't be discouraged if you still see signs of bed bugs after completing these steps. Remember that you have the upper hand now that you have a methodical treatment system, and you are several steps closer to getting rid of them for good!

Below is a list of the important items you will need before you begin the battle. I'll explain where and how to use them in subsequent chapters. I encourage you to read through the entire guide before you start any steps so that you can get a complete picture of what needs to be done. In the process, you might realize that you don't need some of the optional items in the list below, depending on your situation.

When you read through this list for the first time, it may seem like a lot of stuff. However, I would like to point out that the cost of other bed bug treatment systems starts at $1,000 to $2,000 and can climb as high as $6,000 to $7,000 dollars! **Besides, even if you hired a pest control operator to treat your bed bugs, it would still be your responsibility to handle most of preparation and supplemental treatments**. So you would likely incur most of these costs even if you did hire an exterminator.

I will be referring to all of these items below throughout this book, and they will all be discussed again in more detail in the *Bed Bug Supply Manual* that was included with your purchase of this book.

In the *Bed Bug Supply Manual,* you'll learn more about each of the items I have used and where you can get everything quickly and inexpensively. You'll likely have some of these items in your home already. Others will be readily available at local stores. A few of them may need to be ordered. The *Bed Bug Supply Manual* will help you figure all of that out.

Now I know that we agreed on not using any half measures, but there are frugal alternatives to some of these supplies if you find yourself in a situation where you can't afford anything else. You will find a detailed explanation about those alternatives in the free *Bed Bug Supply Manual*. I don't recommend using all of these alternatives but I still provide them because it is always better to resort to a frugal and slightly less effective solution than to skip a portion of the treatment altogether. I believe that a lack of money should never stop you from getting rid of bed bugs so I do my best to provide solutions that are available for everyone. You will get a better idea of which alternatives you can get away with and which ones you should try to avoid in the *Bed Bug Supply Manual*.

Supply List

- Thick garbage bags (contractor bags)
- Re-sealable plastic storage bags, such as Ziploc® bags, in the gallon and 2.5-gallon sizes.
- Vacuum cleaner with hose
- Bucket
- Dish detergent
- Sponge
- Lysol® or other heavy cleaning product
- A mattress encasement for every box spring and mattress in the affected area

- Pillow encasements
- Caulk gun and a few tubes of paintable latex caulk
- Bed bug climb up interceptors
- Joint compound
- Wood filler (if needed)
- Expanding foam (if needed)
- Utility knife
- Rags or paper towels
- Two 4-ounce bottles of CimeXa or any other 100% amorphous silica gel desiccant dust
- Pest control duster
- Either 3-4 rolls of transparent packing tape or 3-4 rolls of transparent adhesive tape with a completely smooth surface, such as Scotch™ tape. (The packing tape is a slightly better option, but either option will work.)
- Steam cleaner
- White sheets, white pillow cases, and white cotton blankets. (Cotton sheets and cotton blankets are easily washed and dried. Comforters may harbor bed bugs even after a long stint in the dryer.)
- Duct tape
- Murphy's Oil Soap (for wooden bedframes), which is a contact killer for bed bugs and is good for cleaning wood and rendering it free of bed bugs

Optional Items
(You'll learn more about these later.)

- Mineral oil

- Vaseline® or other petroleum jelly
- Pesticide
- Dissolvable laundry bags

Chapter 7: The Importance of Containment

The next step is to protect your bed. This guide will instruct you how to do that, and doing so will provide you and your family with some immediate relief from bed bug bites.

Many pest control professionals will argue that you should leave the bed completely unprotected and allow the bed bugs to continue to eat you alive. The theory behind that is they want to keep the infestation concentrated and prevent the insects from fleeing to other sections of your home. That's easy for them to say! They aren't the ones suffering from sleepless nights with no relief in sight.

This is where this bed bug treatment plan differs from every other treatment plan I have ever come across. I am not completely convinced that bed bugs will flee to other areas of your house just because you protect your bed. However, the risk still exists that they may try to flee the treatment area once we begin other pest control procedures, so we want to take a few precautions to limit the infestation to its current area.

There are two important keys to this step. The first is to make sure the bed bugs are trapped or killed during their attempts to climb onto your bed. The second is to take measures to make sure any bed bugs attempting to flee will come into contact with one of your other treatment methods.

Bed bugs are very small and easily capable of moving from one room to another though tiny cracks and crevices. We need to eliminate the pathways they could use to enter and exit a room. If you live in a dwelling that attaches to other units, such as an apartment or townhouse, it is extremely important that you cut off their movement to and from the other units. These adjoining dwellings put you at a higher risk of becoming infested or even re-infested after a successful treatment.

Here are instructions to secure your home and contain an active infestation.

Items You Will Need

- A few rolls of transparent packing tape with a completely smooth surface. Alternatively, you can use transparent adhesive tape such as Scotch® tape, the shiny kind, with a completely smooth surface.
- CimeXa (silica gel desiccant dust). I used diatomaceous earth successfully for many years but have since made the switch to using 100% amorphous silica gel desiccant dust. It does all of the things that diatomaceous earth does, only better and more quickly. It is equally safe and inexpensive.
- Pest control duster. You may need to order this online. I was unable to find one at any local stores. However, you could also use an empty squeeze bottle if you can't get your hands on the duster. It won't work quite as well but it will allow you to get the job done without delay. Things will get a bit messier if you're using a squeeze bottle instead of a pest control duster, so you should wear a mask and a pair of goggles during application if you go that route.

A Quick Word About 100% Amorphous Silica Gel Desiccant Dust and Why I Prefer It over Diatomaceous Earth

Sometimes referred to as "ASG dust," "100% ASG dust," "desiccant dust," or by the brand name "CimeXa," 100% amorphous silica gel desiccant dust is the MVP of our treatment plan. Hereafter, I will mostly refer to it as "ASG dust."

ASG dust is synthetically produced from sand using a simple manufacturing process. Despite the phrase "silica gel" in its name, the material is not actually a gel but is a hard substance that is usually formed into beads, granules, or dust. The word "amorphous" refers to its being non-crystalline, which is the quality that makes it safe for humans and animals. Synthetic forms of amorphous silica gel are commonly used as drying and anti-clumping agents in powdered foods, pharmaceuticals, and cosmetics.

ASG dust works better than diatomaceous earth because it has a static cling effect that helps it to adhere to a bug's body more effectively than diatomaceous earth does. You will want to apply the dust in very thin layers so that bed bugs are not deterred from coming into contact with it. If the dust is applied too thickly, they will simply avoid it, and you want them to come into contact with it. Once it attaches to their bodies, it quickly absorbs their protective coating and kills them within twenty-four to forty-eight hours.

In addition to being a far superior bed bug killer, ASG dust is also a safer and more practical option than diatomaceous earth. For instance, it can be safely applied to open areas, while diatomaceous earth is approved only for crack and crevice treatment. Because ASG dust generally contains a minuscule amount of crystallized silica, it does not pose an inhalation danger or risk of damage to your lungs as does diatomaceous earth. You will read more about ASG dust in a later chapter.

Create a Bed Bug Barrier around Infested Rooms

It's time to get started.

Your first goal is to create a barrier around each room that your inspection showed was infested with bed bugs. You will use ASG dust and smooth Scotch tape or clear packing tape to create this barrier.

1. Apply the dust around the borders of the room where the carpet edge or floor meets the wall. Usually you will want

only a super thin dusting of this stuff. Since the main purpose of this step is to provide a containment barrier, you can go a little bit heavier to ensure all areas have adequate coverage. However, it is very important at every stage that you don't apply too much, because, as stated above, you want as many bugs as possible to come into contact with the dust and not merely avoid it.

Eventually, you'll want to get a super thin dusting inside every crack and crevice you can find as well, and this guide will give you more information about that in a later chapter. For now, just focus on complete room containment, which you will achieve by "building a wall" around the borders of the room.

I highly recommend getting a proper pest control duster to apply the ASG dust. A pest control duster is relatively inexpensive, does a much neater job than a squeeze bottle does, and allows you to apply the dust in hard-to-reach places. In a pinch, I have used squeeze bottles, but if you use them you will encounter some trouble with clogging, applying the proper amount, getting adequate coverage, and trying to distribute the dust into hard-to-reach areas. Save yourself some frustration and the risk of doing a less than full-measure job by investing in a proper pest control duster.

2. Use a small painter's brush to apply a thin layer of ASG dust around doors and to other areas that are not practical to treat with the duster. This stuff has a static cling so it will attach to some vertical surfaces. You want to try to prevent bed bugs from using the center of the walls or ceiling to move from one room to another through the middle or top of doorways.
3. Rather than caulking or sealing cracks, crevices, and voids in the wall and perimeters of the room at this stage, apply a light coat of ASG dust. We don't mind if the bed bugs

continue to harbor in these areas for a little while longer because it reduces the likelihood that they will move around and infest other areas or belongings.

4. Apply a horizontal layer of very smooth clear packing tape or Scotch tape along the wall, 6-12 inches above the baseboard. Bed bugs have claws that can't grip to smooth surfaces, so the tape will act as another containment barrier. You'll need to go around the entire room and be sure there are no gaps. Check to see how many feet of tape are included in each package so you know how much you'll need to cover the entire room. Try to cover as much wall as possible without tearing the tape and starting with a new piece. It helps to have two people working together for this step.

   When you reach a doorway, make a 90-degree turn and go vertical with the tape around the entire door. This will keep bed bugs from using the ceiling or high parts of the wall to crawl through the top of the doorway. You can also dust the trim on the top of the door with ASG dust for extra protection.

   Optional: You can also repeat this process with a horizontal layer around the room up near the point where the wall meets the ceiling. It's not common, but there have been documented cases of bed bugs dropping down from the ceiling onto beds.

5. Create this same barrier in and around any adjoining rooms that could also be infested. Every infestation is different so this will depend on what you found during your inspection. Some people can get away with creating a barrier around only one room. Others may be required to create a barrier around every room in the house.

Chapter 9: Should You "Protect" or "Isolate" Your Bed?

When it comes to getting rid of bed bugs, one of the hottest debates among pest control professionals is bed protection versus bed isolation.

What Is Bed Protection?

Protecting your bed means that you make sure bed bugs are not harboring in the bedframe, headboard, or mattress. To protect your bed, you will encase mattresses and box spring in high quality bed bug-proof encasements. Bed bugs will still be able to crawl onto the bed and bite you, but you are taking steps to ensure they will not be able to live in your bed. Perhaps you will derive some satisfaction from knowing that when they cross treated barriers on their way to you, any meal they enjoy will likely be their last.

Now, protecting the bed is going to sound less appealing than isolating it because you won't be stopping them from biting you altogether. You will drastically decrease the bites, but they will still have access to you. However, there are some benefits to gritting it out this way, and one of the chief ones is that you can have some certainty with bed protection that once the bugs are gone, they are well and truly gone.

What Is Bed Isolation?

Isolating your bed means you are completing all of the protection steps to get them out of your bed and then taking additional steps to make sure they can't get back into it. This option is controversial among some pest control professionals. Some experts believe that it is a bad idea to completely isolate the bed because bed bugs will disperse and spread further throughout your dwelling. Other professionals recommend that you do isolate.

Which Method Should You Choose?

What do I think? I think the debate will rage on for years to come.

I think that if you can deal with a few more nights of being bitten, then I recommend that you follow the instructions for protecting your bed, not isolating your bed.

The main advantage to bed protection over bed isolation is that having bed bugs biting you in bed or finding evidence they were there are both sure signs that you still have bed bugs and your home requires further treatment. On the other hand, if you isolate your bed and your body doesn't react to bites you get during the day, it may be more difficult to verify if bed bugs are still present, allowing a resurgence of an infestation you may have thought was gone.

If you have followed the steps in this book to this point, you have already established strong containment around the borders of the room (see Chapter 7), and you will also be applying more treatments that I will describe in the next chapter. So I don't think that whether you isolate your bed or not will make or break the process.

However, a strong argument can be made that the ability to easily detect and monitor the presence of bed bugs throughout the following treatment stages—which bed protection allows you to do, and bed isolation does not—will lead to a quicker and more efficient removal of the infestation. If on the initial treatments you happen to miss a few of these elusive creatures, you'll be able to detect them easily and prevent the infestation from coming back if you have protected your bed rather than isolated it. With bed isolation, the bugs will immediately stop biting you while you sleep, but you might also lose sleep because, filled with anxiety, you can't confirm whether they are truly gone or quietly making a comeback.

Why Would You Choose to Isolate Your Bed?

If you are being bitten very badly, or you have severe allergic reactions, distress, or mental anguish, you can choose to isolate your bed without fear of jeopardizing the entire treatment. If you do decide to go this route, you will need to make doubly sure to keep the containment areas strong. Do that and you won't have to worry about bed bugs spreading throughout your home. Just make sure the ASG dust is spread properly around the borders of the room so that bed bugs can make contact with it at all points. As long you maintain the containment areas, the bed bugs will have nowhere to run and nowhere to hide.

Chapter 10: Treating the Bed

Whether you have decided to protect your bed or isolate your bed, the next step is the same, as both methods require you to eliminate bed bugs from the mattress, box spring, headboard, and bedframe, and then encase the mattress and box spring.

Everyone will need:

- Encasements for mattress, box spring, and pillows
- New pillows
- White sheets, white pillow cases, and white cotton blankets. Cotton sheets and cotton blankets are easy to wash and dry, but comforters may harbor bed bugs even after a long stint in the dryer.
- Quality duct tape
- Thick contractor bags
- Re-sealable plastic storage bags, such as Ziploc bags, in the gallon and 2.5-gallon sizes
- Murphy's Oil Soap (for wooden bedframes). Murphy's is an effective contact killer for bed bugs and is also useful for sanitizing wood.
- ASG dust (also referred to as 100% amorphous silica gel desiccant dust or by CimeXa, a brand name) See Chapter 13 for more about ASG dust.
- Steam cleaner

If you are isolating your bed, you will also need:

- Bed risers, to elevate the bed and keep bedding from touching the floor
- Mineral oil

- Vaseline
- 4 sturdy bowls or bed bug climb up interceptors for holding mineral oil under the legs of the bedframe.

For everyone, whether you are protecting or isolating your bed:

1. Strip the bed. Put all of the dirty linens into a garbage bag and tie it off tightly. Make sure the bag is airtight by pushing on it to see if it deflates.

2. Launder your bedding as soon as you can. Use extreme caution when opening the bags so that you don't expose your laundry room to bed bugs. In the *Bed Bug Supply Manual*, I'll show you where you can get bags that go straight into the washer and safely dissolve with hot water.

3. Wash in hot water and dry on high heat for the longest possible cycle. Heavy items may require two complete cycles on high heat to ensure everything has been exposed to lethal temperatures for a sufficient amount of time. When you take the items out of the dryer, put them immediately into a new bag to protect them from re-infestation.

4. Repeat these steps with any other fabrics, curtains, and clothing in the infestation areas after you are finished treating the bed.

5. Move the bedframe away from the wall.

6. Thoroughly vacuum the mattress and box spring, keeping in mind that the area in your home that is most susceptible to bed bug infestations is the box spring.

Before you begin, read Chapter 11 for instructions on how to properly use a vacuum cleaner to eliminate bed bugs and their eggs.

Vacuum slowly and methodically along the edges, folds, seams, and surfaces of both the box spring and the mattress.

Remove the dust cover on the bottom of the box spring to expose the framing, as that is one of the bugs' favorite places to hide. Carefully vacuum every inch, crack, and crevice inside of this area to remove as many bed bugs as possible. Keep a close lookout for eggs during this entire process. This is where a UV light will prove extremely useful, since eggs can be difficult to spot due to their size and transparency. The suction of a standard vacuum cleaner is usually not strong enough to remove all of the eggs on its own since they are coated in a sticky cement-like substance. You'll need to scrape them with the vacuum nozzle or secondary brush or scraping tool.

7. Vacuum the bedframe and headboard. Since the bedframe and headboard will not be encased, you'll need to pay special attention to these areas. The bedframe in particular can be a difficult item to treat as there are usually numerous cracks and crevices where the critters can hide. Follow the same steps as given above, and vacuum along every crack and crevice. Depending on the type of bedframe you have, it may even be necessary for you to take it apart so you can have full access to important areas. Treat all crevices with ASG dust or steam to kill any remaining bed bugs and eggs that you were unable to vacuum.

8. Use wood filler or caulk to seal the cracks and crevices you just treated. You can even apply it into screw heads and other openings that are necessary for assembling and disassembling the bed. This could prove troublesome if you have to take your bed apart at some point, but that was the least of my concerns when I was treating my home.

9. Vacuum under and around the bedframe thoroughly. Your main goal is the physical removal or destruction of as many bugs and eggs as possible before you install the encasements.

Pay close attention to seams, mattress ticking, and any voids that are difficult to see into.

10. For extremely heavy infestations, you can apply ASG dust or a round of steam to each section of the bed components to kill any remaining bed bugs. This is an optional step, so if the bed looks relatively clean after vacuuming, you can skip this step and move on to Step 11.

    - If you decide to steam, first read the upcoming section in this manual (Chapter 12) about how to properly use a steamer to kill bed bugs.

    - To prevent moldy conditions, allow the mattress and box spring to dry completely before you enclose them in the encasements.

11. Cover your mattress and box spring in the mattress encasements after they have been treated. There are many encasements available in the marketplace, but few are effective against bed bugs. I'll give you a list of encasements that are effective against bed bugs in the *Bed Bug Supply Manual*. The encasements can be used to salvage infested beds or to protect new ones. Once they have been installed, any bugs or eggs that are still inside will be trapped inside the encasement and eventually die.

12. Put the mattress and box spring back on the frame carefully, so you don't rip the encasements. Encasements are effective only if they are completely intact, so it is important to periodically inspect them to be sure that they have not somehow become ripped, torn, or worn through. Areas where the encasement comes in contact with sharp edges or protrusions, such as bolts or the bedframe, are particularly susceptible to becoming damaged. Placement of some type of padding, such as duct tape, over these areas may help increase the longevity of the encasement.

13. If you chose to protect your bed rather than to isolate it, apply a very thin dusting of ASG dust around the bed posts, but use the ASG dust conservatively. Unless you are isolating your bed, you don't want to merely deter the bed bugs, you want them to come into contact with the dust. Go super light with the dusting or you will end up with an "isolated" bed.

If you are isolating your bed, complete the following steps:

14. Put your bed on risers to ensure that your sheets, blankets, and other bedding do not touch the floor while you sleep.

15. Place sturdy bowls under the legs of the bed and pour mineral oil into the bowls. Any bed bugs trying to climb onto the bed will become trapped in the oil.

16. Wrap cellophane tape around the legs, and coat above and below the tape with petroleum jelly (Vaseline). Both the tape and Vaseline can get dusty and lose their ability to trap bugs, so you'll need to replace them periodically.

17. Vacuum again to pick up any strays that fell or crawled off the mattress and box spring in the process.

18. Remember that you yourself can carry bed bugs into the bed with you, so wear only clean clothing which has been washed, dried, and stored in a location that is safe from bed bugs when you get into bed. To keep your clean clothing safe from bed bugs, you can store it in a tightly sealed contractor bag, re-sealable plastic bags such as a Ziploc bags, or storage bins. No ASG dust is required for storing clothing.

Everyone should complete steps 19-25:

19. Break out the new pillows and put pillow encasements on them.

20. Put clean white linens on the bed. Clean white linens will enable you to see bed bugs, blood, or other stains more easily if any still remain after your treatments.

21. Change and launder bedding every three to five days. Check the sheets every day for bugs, molted skins, blood spots, and stains that look like spots of black ink. If you protected your bed, this is evidence that you still have bed bugs in your home. This can be helpful information, especially if you do not react to bites. If you isolated your bed, this is evidence that the bugs are still in your bed.

22. If you find evidence of bed bugs, consider repeating the steps above for vacuuming and cleaning the mattress, box spring, and bedframe, to ensure bed bugs are not living in your bed. In any case, continue to repeat all treatments approximately every two weeks until you no longer see signs of bed bug bites and all other evidence.

23. Take precautions to make sure the bed bugs and the eggs that were collected in the vacuum cleaner are handled safely to avoid spreading them around your home. Read more about proper vacuuming and handling of the vacuum cleaner in Chapter 11.

24. Examine all of your precautions daily. Maintain your containment areas. Inspect your encasements for holes or tears.

25. Keep all pets away from the infestation area.

Additional Considerations

Wooden Bedframes: If you have a wooden bedframe, take it completely apart and clean every inch of it with Murphy's Oil Soap. Spray the Murphy's Oil directly onto the bedframe components (not onto a rag first). Pay close attention to all the little cracks and crevices in the wood and in the joints. Murphy's Oil Soap kills bed bugs on contact, but it provides no residual protection. Apply a thin layer of silica dust when the oil has dried.

Upholstered Headboards and Footboards: Any upholstery is very difficult to treat successfully. Remove upholstered headboards and footboards from the frame and treat them with a vacuum and ASG dust.

Can I Just Throw My Bed Away?

I understand the level of disgust that can lead to this question. However, it's best not to throw your bed away for a number of reasons.

First off, bed bug infestations are not limited to the bed. Throwing away your mattress and box spring will not solve your problem. New mattresses brought into the home will quickly be infested, and you will be right back where you started.

Second, there is a high probability that bed bugs and eggs could be dispersed throughout your home during the removal process if the mattress is not handled properly.

Finally, we also need to consider where that mattress is going when you throw it away. How long will it be sitting there? Is there a chance others will accidentally come into contact with it? While disposing of your bed might sound like a good idea, it won't do you any good to spread the infestation through the neighborhood because it will eventually find its way back to you.

So, the bottom line is that throwing away your mattress will not solve your bed bug problem.

With that fairly strong caveat in place, you might still have reasons to throw your mattress, box spring and bedding away. **However, you must take care to do so properly.** If your bed is severely infested and needs replacing anyway, you can eliminate a large portion of the infestation in one fell swoop, but don't think it will cut down on your workload. It won't. Getting rid of your mattress, box spring, and bedding properly will take about as much work as treating and protecting your bed. However, you might gain some peace and mind, especially if you're dealing with a heavy infestation.

If you decide to throw your bed away, you MUST follow these instructions to protect yourself, your neighbors, and trash collectors:

- Meticulously wrap items you are discarding in plastic or shrink wrap. Use duct tape to keep the wrapping from coming unraveled.

- The infested items should be handled with extreme caution. Avoid shaking, dropping, or banging them into walls.

- Mark infested items with red spray paint to let others know they are not safe to take.

- Coordinate the disposal of infested items with your neighborhood trash collection schedule. The less time these items have to spend outside, the better.

Remember that bed bug populations will not be limited to the items you are throwing away, and new mattresses can be infested quickly. If you bring a new bed into your home, make sure you protect it by using encasements for both the mattress and the box spring. Only when they have been protected by a proper bed bug encasement, will they be completely safe to bring into an infested room. Bed bugs may still make their way onto encasements, but their access will be limited to the exterior. Just be careful that the encasement is not punctured or ripped during delivery and installation onto the bedframe.

Chapter 11: Vacuuming Instructions

Vacuuming certain areas of your infestation such as the bed is a very effective way to reduce large numbers of bed bugs quickly, and it will also provide immediate bite reduction for bed bug victims. It is important to remember, though, that while vacuuming is an important weapon in your arsenal for fighting bed bugs, you cannot make it your only weapon.

While most types of vacuum cleaners are suitable for this stage of treatment, I have found that vacuum cleaners with bags and a hose attachment are the most convenient, as they make cleanup a bit easier. However, many people have used canister vacuums successfully, too. Just make sure the vacuum you choose comes equipped with a "high efficiency particulate arrester" (HEPA) filtration system.

Vacuum cleaners and attachments to avoid:

- Battery-charged or handheld vacuum cleaners often don't have enough suction power to dislodge bed bugs and their eggs.
- Brush hose attachments with bristles will work to the bugs' advantage, not yours. Bed bugs and their eggs tend to cling to such surfaces. Use a scrub brush that you throw away afterward if you need to scrape any surfaces.

How to Handle Your Vacuum Cleaner When You're Using It to Remove Bed Bugs

You have to be careful how you handle your vacuum cleaner when you are using it to remove bed bugs, as vacuum cleaners can become infested. If at all possible, it is always a little safer to use a separate vacuum cleaner from the one you use routinely in your home. However, this is not always a practical or available option. Here are the steps I recommend you take to reduce the likelihood that your vacuum cleaner ends up harboring bed bugs.

1. Sprinkle a small amount of ASG dust into your vacuum bag or canister before you start removing bed bugs. This will help to create more inhabitable conditions for the varmints.

2. When you have finished vacuuming, seal the vacuum cleaner in a contractor bag and remove it from your home.

3. Once you have it safely outside, carefully empty the contents of the canister or vacuum bag into a separate contractor bag. Seal that bag tightly and insert it into a garbage bin with a tightly fitting lid to avoid the possibility that anyone else will come into contact with it.

4. If you are using a canister vacuum, the canister and filter should be washed thoroughly.

5. You can treat the bottom of the vacuum where the rotating bristles are located with steam to kill any bed bugs and eggs that may be clinging there.

6. When you have finished vacuuming, sprinkle a little more ASG dust into the vacuum bag, canister, and other parts of the vacuum that might have been exposed to bed bugs.

7. Another option is to seal the vacuum in an airtight bag along with a Nuvan® ProStrip®. This product contains a pesticide that is released from the strip as an invisible vapor that is lethal to all life stages of bed bugs, including their eggs. You will read a little bit more about Nuvan ProStrips in an upcoming chapter. Just be aware that the ProStrips take 7-10 days to be completely effective, so if you choose this option, you will be without this vacuum cleaner for a little while.

Vacuuming Techniques

- Use a flashlight or a UV light source if one is available to you. UV flashlights are inexpensive and are useful throughout the entire process of removing bed bugs from your home.

- A magnifying glass is also helpful because bed bugs can hide just about anywhere. Areas as small as the crevice in the head of a screw can harbor bed bugs.
- Move the nozzle along infested areas slowly.
- Use the tip of the nozzle or a throw-away scraper or brush to dislodge eggs.
- In addition to vacuuming the mattress, box spring, headboard, and bedframe as laid out in the previous chapter, you should also apply these vacuuming techniques to any other fabrics, carpets, furniture, and anything else you can think of. The more items you vacuum, the more bed bugs you will remove.
- Bed bugs tend to gather and hide in the carpet edges where the silica dust was applied for containment. While you probably will want to vacuum these areas, it's important to maintain the integrity of the containment. I like to vacuum up two or three feet of the containment barrier at a time and then reapply the dust to maintain containment before I move on to the next section. This is especially important with heavy infestations.
- Remember that early in this book I warned you not to start vacuuming the bugs up without first following the other instructions in this book? It seems like a lot of work to do things in order as I have described—and it is—but if you start vacuuming those carpet edges before applying the ASG dust to them, and then reapplying ASG dust immediately after vacuuming, the bed bugs are going to scatter all over the place without coming into contact with treatments. We need to avoid that at all costs.
- A vacuum cleaner is a critical tool in the bed bug elimination process, but you will never be able to remove the entire

infestation with only one treatment method. As good as it will feel to physically remove so many of these critters, it is important that you don't let your guard down. You are gaining the upper hand, but you can't stop here!

Chapter 12: Steam Instructions

Steam is another effective treatment method for eliminating bed bugs that is used by almost every pest control professional. Steam differs from ASG dust and vacuuming in that if it's done properly and with the right equipment it instantly kills bed bugs in all of their life stages on contact. However, if you choose to use steam, do so after vacuuming but before applying ASG dust, as steam will likely neutralize the killing effects of the dust.

When you are steaming, avoid the carpet edges. While it's true that the bugs love to hide there, it's best just to vacuum the carpet edges, not steam them, and reapply the ASG dust immediately. This is because it is so important to maintain the integrity of the room containment (see Chapter 7) to prevent the bugs from moving into other parts of your home. Room containment is a critical early step that should always be established before treating anything else—and maintained throughout the other treatment phases—to keep the infestation from spreading.

If you need a refresher about how best to vacuum the carpet edges see "Vacuuming Techniques" in Chapter 11.

Some Advice About Using Steam

- Steam is dangerous. Follow the manufacturer's instructions carefully and always stay focused on what you are doing.

- Make sure you use a dry vapor steam unit to reduce the risk of mold. Regular steamers will leave excess moisture in everything you treat.

- You can steam almost anything in your house, including floorboards, furniture, baseboards, walls, ceilings, nooks, and crannies.

- However, never apply steam to electrical outlets, wires, or electronic devices.

- As mentioned above, avoid steaming the carpet edges.
- Steam must hit bed bugs directly but not so hard that it blows them away rather than killing them.
- Temperature is the most important factor, since heat is the bed bug's Achilles heel. The temperature at the tip of the steamer as listed by the manufacturer should be at least 200 degrees Fahrenheit. That will leave steamed surfaces at a temperature of about 155-165 degrees, which will kill all bed bugs and their eggs instantly. In the *Bed Bug Supply Manual*, I give recommendations about finding a steamer that will do the job.

Key Things to Look for in a Steamer
- Dry or "dry vapor" steam (This is important.)
- Heat at the tip of at least 200F degrees per the manufacturer
- Large water chamber to cut down on starts and stops in the steaming process
- A decent length of cord (or an extension cord)
- A good warranty on the boiler

There are many steamers on the market, but not all of them are useful for killing bugs. The steamer I use is the Vapamore MR-100, as it is the most affordable dry vapor steamer that meets all of the key criteria on the list above. It currently retails for about $300-$350. You can find out more about it and learn where to get it for the best price in the *Bed Bug Supply Manual*.

You can also call around to see if you can rent a steamer from a local tool or equipment rental business. The closest one to me that had a good steamer was about forty minutes away from my home, so I opted to buy my steamer instead. If you decide to rent, you should be able to rent one for about $50-$100 for a one- or two-day rental.

Application of Steam

Always use extreme caution when using a steamer, because the high temperatures can cause severe burns. Move slowly and deliberately so that all of the bed bugs and their eggs that the steam comes into contact with are killed. With the proper temperatures, you can expect to move about one inch per second. If the steam is coming out too fast, fit a thin cloth over the nozzle. This will decrease the force of the steam without affecting its temperature.

Chapter 13: 100% Amorphous Silica Gel Desiccant Dust (ASG Dust)

We have already touched on 100% amorphous silica gel desiccant dust (also known by several other names, including ASG dust and the brand name CimeXa) in previous chapters. In this chapter, I want to give you a deeper look at this material so you can fully understand how to use it properly.

What We Know Already

- I believe that 100% amorphous silica gel desiccant dust (ASG dust) is the most effective material for killing bed bugs.
- It may also be referred to as "ASG," "silica dust," "CimeXa," or "desiccant dust."
- It has a static cling effect that causes it to adhere to bed bugs' bodies.
- Once bugs have come into contact with it, ASG dust quickly absorbs a protective layer that holds moisture in bed bugs' bodies, thus killing them within twenty-four to forty-eight hours of first contact.
- Its static cling effect provides far more exposure than other materials and leads to an accelerated kill as compared with diatomaceous earth or any other substance available.
- It is a very safe product and one of the only bed bug products that is rated for use in open areas, including in and around the bed. Most of the other products on the market are hazardous to humans in one way or another and are rated only for crack and crevice treatment.

How Safe Is It?

ASG dust is safe enough so that it is commonly used in or packaged with foods and cosmetics. On the label for CimeXa, a brand name for ASG dust, is a statement that the contents are intended to be used on bare mattresses, beds, carpets, baseboards, and so on. This statement illustrates just how safe and versatile ASG dust is, and because it is also amazingly effective in killing bed bugs, ASG dust is as close to a perfect weapon against bed bugs as you will find.

How Should I Use It?

Treat as many surfaces in the infested area as possible to maximize the chances bed bugs will come into contact with ASG dust. This includes all baseboards, trim, door and window frames, holes in the walls, damaged wall paper, electrical switches, and electrical outlets. In addition, you want to dust holes for plumbing, electricity, heating and air conditioning equipment, and any other wall or ceiling fixtures. Keep in mind that if the crevice is large enough to fit the edge of a credit card into it, it is large enough for a bed bug to hide in or pass through.

ASG dust will adhere to some vertical surfaces but you cannot rely on it to be effective in those areas.

As I mentioned earlier, it is always important to dust lightly. If any desiccant dust is spread too heavily, bed bugs will avoid it altogether instead of coming close enough to make contact. This is why it is so important to use a proper applicator. See the *Bed Bug Supply Manual* for details.

With room containment, vacuuming, and proper application of ASG dust, you can easily terminate most of the infestation in your home within the first few days.

Chapter 14: Organization, Containment, and Treatment of Belongings

Now comes the long and somewhat tedious process of sorting and storing personal items into bags and containers. While this piece of the process will take some time, it is critical.

However, I want to encourage you not to lose heart now, not after you have come so far. It is more important than ever to keep on keeping on. Otherwise, all your previous efforts will come to naught.

As you work through this phase, label each bag or container with its contents and whether those contents are "possibly infested" or "safe." Treatment methods will vary depending on what is in each bag, so keep similar items grouped together.

Step 1: Organize, bag, and treat washables such as clothes, curtains, shoes, bedding, stuffed toys, pillows, etc. Wash these items in hot water and dry them on the highest heat setting possible. Heat is the key element, so if clothes are already clean, they don't necessarily have to be rewashed. Just put them in the dryer on high heat for thirty minutes. However, since I know that bed bugs can also drown, I like to put nearly everything through both the washer and dryer.

Step2: Organize, bag, and treat hard items such as electronics, books, paintings, jewelry, and any delicate items that can't be washed. These are more challenging items to treat. Some of these items may dusted with ASG dust, but use your best discretion. Most of these things will need to be isolated and sealed in plastic bags, such as Ziploc or contractor bags. Once these items are isolated in bags you have some options about how you can kill the bed bugs that are inside the bags:

Your options are:

1. If you want to keep the items but have no immediate use for them, you could simply starve the bed bugs to death. However, bed bugs can live for over a year without feeding, so be prepared to lock your items away for at least eighteen months.

2. Insert a Nuvan ProStrip into the bag with the items and reseal it. When you choose this method, be careful not to overstuff the bags, as I believe the ProStrips are more effective when there is plenty of airspace between the items in the bag. Also, be absolutely certain that the bag has an airtight seal, and read the manufacturer's instructions and follow them to the letter.

    ProStrips contain a pesticide that releases from the strip as an odorless, invisible vapor that is lethal to all life stages of bed bugs, including eggs. The vapors emerge from these strips slowly over time, so you should leave items sealed in the bags with them for at least a week to ten days. It takes longer for the ProStrips to kill the eggs than it does to kill the hatched bed bugs, and younger bed bugs are killed more quickly than adult bed bugs.

3. Portable heaters are available to treat smaller items in your home. This is a more expensive option but they are extremely effective. They range in cost, depending on size, from $200 to $350 dollars. Portable bed bug heaters are available in various sizes and kill 100% of bed bugs in all life stages, including eggs. They are very useful for harder to treat items such as books, files, electronics and other household items. These heaters can also be utilized to treat luggage upon return from travelling to ensure bed bugs are not transferred into your home. You'll find more information portable bed bug heaters in the *Bed Bug Supply Manual*.

4. Bed bugs and their eggs need oxygen to live, so you could also vacuum seal the bags to kill the bugs by asphyxiation. If you choose this method, add dry ice to the bags before you vacuum seal them. For small bags containing smaller items such as framed paintings or purses, add one pound of dry ice. Add two pounds of dry ice to medium bags containing items such as suitcases and smaller appliances, and use three pounds of dry ice in large bags containing items such as small furniture pieces and larger entertainment systems. The carbon dioxide (CO_2) from the dry ice will replace any oxygen that happens to be left in the bags. This method takes about forty-eight hours but I suggest you double that just to be safe.

5. Some hard toys and other non-delicate items can be put into mesh bags and run through a dishwasher with "heated dry" turned on. However, I don't recommend this method unless you already know your entire home is infested, since you would be taking a risk by moving infested items to an area of your home that is possibly not infested. If you do choose this method, use extreme caution. I would avoid it unless absolutely necessary.

Step 3: Protect and isolate all washed and treated items in new contractor bags or storage bins so that bed bugs can't infest them again.

Step 4: Throw away any small items that you don't absolutely need anymore—but only if those items can fit into a contractor bag or a large Ziploc bag. You don't need to throw everything away, but if the item is small enough to dispose of safely, without risk spreading bed bugs throughout your home or into other people's homes, then you can get rid of it.

If there are somewhat larger items that you are considering throwing away, such as dressers, night stands, and other items that cannot be safely sealed inside a contractor bag or other plastic bag, remember what you read earlier about the dangers of throwing an infested mattress away. The same rules apply with these items. Therefore, it would be much easier and safer to treat these types of items.

Place anything you decide to throw away into contractor bags and make sure the bags are tied securely to create an airtight seal. Next, either label the bags with the words "bed bugs" or keep them in a secure area where other people will not tamper with them until trash day.

Let's Roll!

By now the infested areas should be completely clutter-free. There should be no clothes or bedding in the room or in the closets, no clocks, no pictures on the wall or elsewhere, no throw rugs on the floor, no junk, debris, knick-knacks, or any other small items on tabletops or anywhere else in the area. Because you've maintained the integrity of your room containment and followed all the steps in order to this point, you've got the bugs right where you want them. Are you ready? Let's go!

Chapter 15: It's Time to Kill Some Bed Bugs!

This is the moment you've been waiting for. Up to now, all your time and efforts have been spent in preparation. Now, let's kill some bugs!

You will need:

- ASG dust
- Pest control duster (You can use a plastic squeeze bottle with a pointed cap or a seasoning shaker if absolutely necessary.)
- Cosmetic brush or paint brush with very fine bristles (for areas where you will need more control over the application)
- Screwdriver to loosen electrical switch plate and outlet covers
- Dust mask
- Vacuum cleaner with a HEPA filter

Instructions

1. Lightly treat the entire floor or carpet with ASG dust.

    Note: You will be keeping a very light dusting of this stuff on the floor or carpet for the next few months. You can vacuum or clean as often as you like and then reapply the ASG dust when you're finished.

2. Pull furniture and other items away from the walls and then take everything apart. All furniture must be turned upside down to allow treatment of all areas, and all drawers must be removed. Drawers should have already been emptied of their contents by now, but if they have not, place everything that is still left inside them into suitable bed bug-proof bags or containers and follow the instructions outlined in Chapter 14.

3. Couches, loveseats, and upholstered chairs will need to be treated in the same manner as you treated the bed if you suspect bed bugs are on them or in them. Remove cushions and dust every crack and crevice possible except for the surfaces that people will sit on. Attention to detail is required here. Use your paint brush and duster to meticulously coat zippers, the undersides of cushions, seams, tucks, folds, buttons, and so forth. Continue to use your furniture the same way you always did. Once you have completed this step, your furniture is now a killing station for bed bugs, and you are the bait. (Sorry!) Your war against bed bugs is almost over!

4. Once a piece of furniture has been treated, you can place bed bug climb up interceptors under the legs to isolate it. (See more about bed bug interceptors in the Bed Bug Supply Manual.)

5. Loosen electrical outlet and switch plate covers. Beg bugs love to hide inside of these. Apply ASG dust to the insides, preferably with a pest control duster, then tighten them again.

6. Apply ASG dust to all remaining cracks, crevices, and voids. Don't neglect to apply the dust behind appliances and other stationary items. ASG dust will remain effective for ten years if left undisturbed, so I recommend applying a thin coating to all non-traffic areas in your home, such as behind cabinets, inside wall voids, around door frames (if there is no caulk under the door trim), and the like. ASG dust will also protect against other common pests such as ants, cockroaches, silverfish, and so on.

7. I don't recommend caulking or sealing cracks, crevices, or voids in your home until you have properly treated them all with ASG dust. We don't mind if bed bugs have access to these areas at this stage because we would rather they come

into contact with ASG dust first. The only exception to this rule is a crack, crevice, or void that would allow the bugs unobstructed access to another area of your home before they have a chance to come into contact with ASG dust. If you feel that is the case, then you can seal that crack, crevice, or void immediately, after giving it a puff of ASG dust. We will revisit caulking and sealing methods later.

8. Some people think of what I'm about to describe as an optional step, but I prefer to do it with *every single infestation* I treat. However, if you rent, check with your landlord before taking this step.

- Use a stud finder to map out a pattern where you will drill holes in the walls in between the studs, just large enough to fit the tip of your duster into.
- If you can, try to drill the holes at a forty-five degree angle right at the point where the top of the baseboard meets the wall. Doing so will allow you to caulk the holes afterward rather than spackling and painting.
- Apply ASG dust into the wall void at each location.

This step will greatly reduce the chances of any bed bugs surviving and will protect your home from future infestations for the next decade. It is an especially important tactic in apartment complexes, condos, townhomes, or duplexes. While this step requires a bit of work to patch and paint the holes, at this point, painting is the least of your worries!

9. Continue applying a very thin coating of ASG dust to all other appropriate surfaces.
10. If you see live bed bugs, castings, or eggs, use a vacuum cleaner, steamer, or ASG dust, depending on the surface area.
11. Maintain the integrity of the containment areas and continue to monitor for bed bug activity.

Can't I Just Use Pesticides to Kill Bed Bugs?

You don't need pesticides to kill bed bugs. Scientific studies have shown that ASG dust is more effective for killing bed bugs than any chemical poison. I have never needed to use a pesticide to eliminate a bed bug infestation.

Chapter 16: Pest-Proofing Rooms

Can you make your home bed bug proof? Yes, you can make your home bed bug proof to a certain degree.

However, doing so will be a somewhat tedious task, and that's likely why this technique is often overlooked. All the same, though, this is not the time to let your guard down. Complete this phase and you will have taken a major step in winning your war against bed bugs, as the technique I describe in this chapter is extremely effective for keeping bed bugs at bay and preventing a re-infestation.

You've come a long way since you first realized your home had a bed bug problem. Now it's time to finish the job.

You will need:

- Caulk gun
- Paintable latex caulk
- Joint compound
- Wood filler (if needed)
- Expanding foam (if needed)
- Utility knife
- Rags or paper towels

The main goal in bed bug-proofing a room or dwelling is to eliminate the bugs' ability to travel from one room to another. Additionally, it is also beneficial to eliminate cracks, crevices, and voids within a room to reduce the number of bed bug harborages. The rule of thumb is that if you can fit the edge of a credit card into a crevice, then it can harbor bed bugs. My additional rule of thumb is that you seal everything you can see.

Common areas that might require attention are cracks and crevices in or on:

- Furniture
- Bedframes and headboards
- Trim work
- Molding
- Door and window frames
- Wallpaper
- Drywall
- Anything else you can see

By now you should have a good idea of the areas that present a problem to the control and elimination of bed bugs. So take these ideas and then use your own creativity to come up with the right bed bug-proofing plan for your home.

Bed Bug-Proofing Tips

Apply a fresh (light) coating of ASG dust to all cracks and crevices you plan to seal. Then use paintable latex caulk, wood filler, expandable foam, or paint to seal everything you can find.

You will probably need a bunch of paper towels, because you can use a damp paper towel to clean up excess caulk and any other mistakes you make along the way.

If you'd like the caulking you do to have a nicer look, there is a nifty little tool you can purchase that will smooth over a bead of caulk to give it a professional-looking finish.

If you have never caulked before, here are a few guidelines:

- Cut the tip of the caulking tube at an angle. The further down you cut, the more caulk will come out of the tube.
- You will also need to pierce the inside of the tube with a pin. There is a seal about three inches down that needs to be

punctured. Most caulk guns have a pin on the side for this purpose.

- For small cracks and crevices you only need a little bit of caulk at a time. Keep in mind that you can always cut more away from the tip of the tube of caulk but you can't add any of the tip back. Therefore, you want to start with the smallest hairline cracks and crevices and work your way up to the larger voids, cutting more of the tip away as you go to allow for larger amounts of caulk to exit the tube.

- Hold the gun at the appropriate angle for each crevice so the flat part of the tip you created by cutting it is open to the crevice you are sealing.

- Squeeze the trigger until caulk begins to exit the tube. Move the tip along the crevice at a slow and steady pace. Sometimes it is more practical to spread the caulk along crevices with your finger to achieve a better seal. You will get the hang of it as you go along.

Messes are inevitable, but they will clean up easily with a damp cloth or paper towel.

Latex caulk is limited to sealing smaller gaps. Larger gaps may require expanding foam or a construction repair.

Chapter 17: Monitor and Maintain

Congratulations! Your hardest work is over!

However, don't let your guard down now. In order to prevent a resurgence, it is important that you maintain the integrity of the ASG dust and monitor for signs of bed bugs.

Routine inspections are your most important monitoring tool. Using the same inspection procedures you followed for your initial inspection, regularly inspect the areas of your home that were infested. If you find any live bed bugs or eggs, vacuum them up or blast them with steam. Make sure you immediately reapply ASG dust to any areas that you clean, vacuum, or disturb.

You can clean anything in the infested area regularly. I like to disinfect because bed bugs are just downright nasty little creatures. So I might clean everything with Murphy's Oil and Lysol and then reapply the ASG dust when I am finished.

Please remember that you cannot rely on the presence (or lack) of bed bug bites to determine if bed bugs are still present. Some people do not have any skin reaction to bed bug bites.

Passive Monitoring Options

Bed bug climb up interceptors can be effective for monitoring. You just need to be certain that the bed or other furniture is pulled away from the wall and that blankets do not touch the floor. The bed posts should be the only way a bed bug can gain access to the bed. Keep in mind that bed bug interceptors will isolate the bed.

You can also use a bed bug trap that relies on CO_2 and heat. You can purchase ready-made traps that work pretty well or you can make your own. In fact, homemade devices work even better sometimes because you can produce larger amounts of CO_2. I'll describe how to do this in the *Bed Bug Supply Manual*.

Your Most Important Monitoring Tool

In the end, though, YOU will be your most valuable monitoring tool. If you inspect regularly and continue to sleep in a "protected" bed with white sheets and a white cotton blanket, monitoring will be much easier.

Chapter 18: Conclusion

So there you have it!

There are two camps in the world: those who have dealt with bed bugs before, and those who haven't.

Battling bed bugs is no easy task, but it can be done. I applaud you for taking the approach that you did with your infestation. The only chance we have to truly win this battle with bed bugs is through education and prevention. The odds are strong that someone you know will also deal with an infestation at some point in their lives. Now you have the knowledge and resources to help them.

I wish you the best.

Frequently Asked Questions

How did I get bed bugs?

It's really hard to say where someone may have gotten bed bugs. You can get them from movie theaters, airplanes, buses, hotels, friends, family, neighbors, used furniture, and other things or places that have been infested. They can also move between dwellings if you live in an apartment or condo complex. Bed bugs could have come from *anywhere*!

Are bed bugs dangerous?

Bed bugs are not known to transmit any diseases. The only physical danger from bed bugs is due to secondary infection caused by scratching and/or neglecting their bites. Scratching bites can lead to a secondary infection. Resist the urge to scratch bites and treat them as you would a mosquito bite. The most severe negative effects of bed bugs are emotional ones. Severe stress, anxiety, and sleep deprivation are common among people whose homes have become infested with bed bugs.

How can I avoid bed bugs when traveling?

Every traveler should learn about bed bugs. Always inspect your hotel room before settling into it. Pack a flashlight (even the keychain LED variety) and gloves to aid in your inspection. The inspection should focus around the bed. If you're traveling alone, ask to have someone from the hotel staff to help you to lift and look under items in the room. Start with the headboard, which is usually held on the wall with brackets. Lift it up one or two inches, then lean the top away from the wall to gain access to the back After checking the headboard, check the sheets and pillows for blood spots. Next, pull back the sheets. Check the piping of the mattress and box spring. Finally, look in and under the drawers of the bedside table. If all these places are clear, enjoy the night. The next morning, look for spots on the sheets. Bed bugs defecate soon after they feed.

If you find evidence of bed bugs but no live ones, the evidence may be old and doesn't mean that the hotel is dirty. Tell the front desk discreetly what you found and ask for another room—one that doesn't share a wall with the room you just vacated. Bed bugs are a public relations nightmare for the hospitality industry. If you run to a competitor (who's just as likely to have bed bugs) it makes it less likely that the industry will become more open about this issue. Communication is key. Ideally, hotels and motels would pride themselves on their bed bug programs and show customers how to inspect to keep all parties bed bug free.

If you can avoid it, don't unpack into drawers, and keep your luggage closed on a luggage rack that's been pulled away from the wall. Never set luggage on the bed.

What can I do if I just got back from a place where there might have been bed bugs?

Launder your clothes before or as soon as these items are brought back into your home. If you found bed bugs after moving into a hotel room, you could ask the hotel to pay for laundering—and for steam-cleaning your luggage. The hotel may refuse, but it's worth asking. Regardless, once home you should unpack on a floor that will allow you to see bed bugs—not on carpets! Unpack directly into plastic bags for taking clothes to the laundry. Suitcases should be carefully inspected and vacuumed, or even freeze them if possible.

Will bed bugs travel on my body?

It is possible but unlikely that a bed bug would travel on you or the clothes you are wearing. You move too much to be a good hiding place. Bed bugs are more likely to be spread via luggage, backpacks, briefcases, mattresses, and used furniture.

What do I do with my pets if I have bed bugs?

The word on the street is that bed bugs will feed only on humans. They certainly prefer human hosts. However, I have spoken to pest control professionals who have seen bed bugs feeding on pets. They won't make their home on a pet like fleas do, but pets could sustain and transport bed bugs. Treat your pet's sleeping area the same way you treat your own.

How long does it take to get rid of bed bugs?

There are a number of factors that influence how long it takes to get rid of bed bugs, including how long the bugs have been present, how quickly you take action, whether other people you live with cooperate with you in your efforts, whether you have made your problem worse by attempting to treat it using ineffective methods, the amount of clutter present, the type and placement of the sleeping furniture, the pesticide resistance level of the bugs, and the physical environment. Here is a little more information about these factors:

- **How long have bed bugs been present?** If activity has just started, chances are good you'll find the bugs in your bed only. (And you might find only one bug if you're lucky!) However, you should still check and treat the whole room to give yourself a good chance to be rid of bed bugs with a single treatment.

- **How quickly did you get into action after you discovered the infestation?** Leave it for too long and you'll give the varmints time to reproduce. However, keep in mind that ineffective methods will only make your problem worse. Most insecticide sprays won't kill bed bug eggs and an insecticide bomb will just spread the vermin deeper into your home.

- **What is the level and extent of the problem?** When infestations have been left untreated for some time, the number of insects could be high, and bed bugs could have found harborages inside multiple rooms.

- **How much clutter is present?** Lots of clutter gives the bugs lots of places to hide.
- **What type of sleeping furniture is in the infested area, and how is it placed in the room?**
- **What is the physical environment of the infested area?** Apartments, condos, townhomes, and duplexes present more challenging circumstances, since bed bugs could be spreading from dwelling to dwelling.

My best guesstimate is that it will take about two or three properly executed treatments over a span of three to five weeks to completely get rid of bed bugs. However, rest assured that you can easily cut the infestation down by eighty-five percent or more in the first twenty-four to seventy-two hours. Some people never see a bed bug again after the first treatment. Every infestation is different so please don't be discouraged if takes you longer.

What if I would rather use pesticides?

I believe that toxic pesticides should only be used in the extermination of bed bugs by a trained pest management professional. Applying toxic substances on your own could make the infestation worse and endanger the people living in your home. Besides, recent studies have shown that ASG dust is more effective for killing bed bugs than pesticides. If you decide to use pesticides for any reason, *please* read and follow the labeled directions.

What if I am a renter?

Renters should not perform any treatments until they report the suspected bed bug infestations to their landlord or building managers. Landlords or property managers should respond promptly. If you don't receive an adequate response from your manager, consider taking things into your own hands by contacting your city's building department for guidance. It's ultimately very important that residents and landlords work together to get rid of bed bugs.

What if I have to move in the middle of a bed bug infestation?

It is not advisable to attempt moving while your residence is infested with bed bugs. If your circumstances are forcing you to relocate, you must take careful precautions to ensure that you don't take bed bugs with you.

You'll need to follow all of the steps in this guide before moving. This will include treating the bed, vacuuming, applying ASG dust as directed, cleaning, and organizing and isolating your belongings. More than half the battle is killing bed bugs that are on your personal items. Leaving the infested dwelling will not allow you to cut any corners.

What if I decide I would rather hire a pest control professional than do it myself?

If you are reading this book, you are probably pretty darn determined to get rid of the bed bugs however you can. And if you've read my story about my own experience with bed bugs, you know that hiring a pest control professional does not offer any guarantee that the results will be better than if you do it yourself. However, I understand that some people just feel more comfortable with the guidance of an experienced technician. Ultimately, I encourage you to follow your heart and do what you think is best for you. No one has more resolve to get bed bugs out of your house than you do. However, if you can find an honest and experienced professional, they can take the pressure off you.

If you can afford to hire a professional, it is critical that you know how to choose the best company. Your area is sure to be filled with good ones and bad ones. The key is to decipher which companies have a proven and successful track record when it comes to getting rid of bed bugs. Take into account the company's pricing, and then do your homework.

Pricing:

- To make accurate comparisons, ask each company what the treatment will cost on an hourly basis per technician. Some companies will use two technicians, and asking about the cost per technician will allow you to compare all quotes equally.

- Overpricing most likely indicates that the company wants nothing to do with treating bed bugs. Sure, they'll take the job if you are willing to pay them a fortune but they would rather avoid it at all costs.

- Underpricing indicates inexperience. Any company that has extensive experience in dealing with bed bugs will not underprice the job.

Do your homework:
- Inspect the company's reviews on sites such as Angie's List, Yelp, and Google.

- Check with the Better Business Bureau to see if any complaints have been filed against the company.

- Ask the company to provide direct references from homeowners they have helped in the past with bed bugs.

- Ask how much they cover in liability insurance. If they do not have liability insurance, hang up the phone.

- Does the company offer any type of guarantee? If so, what is included? Due to the higher chance of re-infestation, most companies will not offer guarantees on apartments or multifamily units. If you occupy one of these dwellings, ask for the company's guarantee on a single family home anyway. A strong guarantee will give you an indication of their level of confidence in the treatment procedures they use.

- Ask the company if they provide a crack and crevice treatment. This is where many bed bugs hide so it is critical that the company has a plan for treating these hard-to-reach areas.

- Ask about how many treatments the price covers. An experienced company will tell you two or three treatments will be required. These treatments should be about two weeks apart, and it will take at least two hours to complete each one.

- Make sure the company does not rely exclusively on pesticides. There are many non-chemical measures that are critical to getting rid of bed bugs permanently.

- Ask if the company employs dedicated bed bug technicians. Experienced companies will usually have dedicated teams for treating bed bug infestations. This is not a deal breaker, but it is a good sign if the company has dedicated teams for treating bed bug infestations.

Bed Bug Supply Manual

Chipp Marshal
Copyright 2016 Chipp Marshal

This manual is a companion to the book *Breaking Bed Bugs* and contains additional information about where and how to obtain all of the supplies you will need to get rid of bed bugs successfully.

There is no shortage of ineffective chemicals, devices, traps, and other methods that are marketed to unsuspecting customers. Bed bugs are big business, and companies are looking to cash in on your desperation to get rid of them. However, just because a product claims to do something does not mean that it will. When you get to the store and see some fancy bottle of bed bug "miracle" spray, do yourself a favor and keep walking.

I have no affiliations with bed bug products or manufacturers. The recommendations I make in this manual and the book it accompanies come from my personal experiences or the experiences of other professionals. I will try to give you multiple brand and manufacturer options when possible so long as it does not compromise quality or effectiveness in the process.

The two factors you must consider when you are gathering materials for bed bug extermination are:

- How quickly can I acquire it?
- How much does it cost?

Your bed bug tolerance and financial situation will dictate which of these factors is most important to you. I will give you the best options for acquiring each product as quickly and/or inexpensively as possible. I will also provide do-it-yourself options where applicable.

It will be impossible to offset the costs for most of these items. As you read through this manual you will undoubtedly be wondering if you can skip, replace, or cut a corner somewhere to save a little more cash. The simple answer is, no. Everything I have ever researched, tested, and written about has been geared towards getting rid of bed bugs as inexpensively as possible without sacrificing effectiveness and success rate. Remember that you would incur these costs and more if you hired a pest control operator (PCO), because you would be charged for each of these items (at their marked-up prices), tasked with doing all the work (aside from pesticide/insecticide application) and then charged thousands of dollars for the PCO's supervision.

100% Amorphous Silica Gel Desiccant Dust (ASG dust)

The most critical part of the treatment strategy I describe is 100% amorphous silica gel desiccant dust (ASG dust). The leading brand of ASG dust is CimeXa, which is manufactured by Rockwell Labs Ltd. I have no ties or affiliations with the company, and I do not get compensated for recommending their product. I have just come to trust it and quite honestly don't know of any other reputable companies that produce a product of similar quality.

Where to Purchase ASG Dust

Unfortunately, ASG dust is not sold in local stores, and you'll need to purchase it online. This is not a perfect situation for someone suffering though a bed bug infestation, but 100% ASG dust is the most essential component of this treatment. Hopefully it will be stocked on the shelves at local big box hardware stores soon, but for now, ordering it online is your only option.

My recommendation would be to purchase it from Amazon.com. They consistently offer the lowest prices as well as providing one- and two-day shipping options, which unfortunately will cost you a little more.

For example, one-day shipping will add an additional $20 to the tab, and two-day shipping costs an additional $12. I don't like paying extra for shipping any more than anyone else does, but we need this stuff and we need it *fast*.

Amazon does offer free shipping on orders over a certain dollar amount. However, that shipping time is going to take seven days on average. I don't recommend waiting that long, but if you don't have any other choice due to financial restrictions, that option is available to you.

Amount Required: Order a minimum of two or three bottles (4ounces each).

ASG Dust Application Tools

At the very least, you will need a paint brush to apply ASG dust. However, I strongly recommend that you buy a pest control hand duster. They only cost between $10-$15 and are worth their weight in gold when it comes to using ASG dust to get rid of bed bugs.

Good hand dusters that I currently own or have used in the past:

- JT Eaton Bulb Duster (about $12)
- Bellows Hand Duster BHD001 (about $12)
- Punchau Bulb Duster (about $28)
- Any other duster with a rating of four stars or more should also work fine

Where to Purchase Hand Dusters

You'll need to grab the duster online when you purchase the CimeXa.

Additional Resources

Here is a link to a helpful video with some tips about using the right tools and how to use them in order to apply CimeXa properly. The gentleman in this video gets a bit carried away with all of the product recommendations such as knee pads and head lamps, but it is a good video to watch for basic information about CimeXA, application tools, and safety.

https://youtu.be/OC8mFqvyH9I

The next video will allow you to see firsthand how CimeXa should be applied to beds, other furniture, and additional items in the infested area.

https://youtu.be/hmZQDmhpiWk

Summary

- Purchase two or three (4oz) bottles of CimeXa and a hand duster with good reviews from Amazon.com.
- Purchase a fine, soft-bristled paint brush from any local paint or hardware store.
- Refer to these videos for tips and a demonstration of a CimeXa application.

Other Items You Will Need

You may already have most of these items at home. If not, they can be purchased locally:

- Thick garbage bags (contractor bags)
- XL and XXL Ziploc® bags
- Bucket
- Dish detergent
- Sponge
- Lysol or other heavy cleaning product
- Joint compound (if drywall filling, repair, or patching is needed)
- Wood filler (if needed)
- Expanding foam (if needed for larger voids that caulk cannot fill)
- Utility knife
- Rags or paper towels
- 3-4 rolls of clear Scotch® tape (or similar tape with a completely smooth surface)
- White sheets, white pillow cases, and white cotton blankets. Cotton sheets and cotton blankets are easier to wash and dry. Comforters may harbor bed bugs even after a long stint in the dryer.
- Duct tape
- Murphy's Oil Soap
- Paintable latex caulk
- Caulk gun

Optional items you will need if you are "isolating" your bed:

- Mineral oil
- Vaseline®

A Quick Tip to Save You Some Time

All of these items can either be found in your home or purchased with a quick trip to your local hardware store. This tip is certainly not exclusive to shopping for bed bug treatment supplies but it is something I recently stumbled upon that makes life much easier. No matter how many times I try to make a quick stop at my local hardware store, I end up spending one to two hours inside looking for everything. A family member who is a general contractor recently revealed to me how he keeps from spending that much time in the store every day.

Rather than wandering around the store for an hour like I always do, you can order all of your items online and select the "pick-up in store" option. It even allows you to select a time to pick everything up. They usually require at least 60 minutes to get your order ready. When you arrive at the store, all of your items will be waiting on a cart for you to take to the register. It doesn't get much better than that!

Just go to the website of your preferred store to get started.

Mattress and Box Spring Encasements

There is no shortage of bed encasements on the market. However, unless the encasement is designed specifically for bed bug protection and has been subjected to rigorous testing, it is likely to be ineffective against bed bugs and/or prone to failure.

Below is a list of mattress and box spring encasements that I have learned are effective against bed bugs. The products on this list are included here based on my own experience, my research, or recommendations from other bed bug professionals:

- Hospitality Sleep Defense Bed Bug Mattress Encasement
- SafeRest Bed Bug Mattress Encasement
- Sleep Guard Premium Mattress Encasement
- Sure Guardian Bed Bug Mattress Encasement

If you want to purchase an encasement that is not on this list, just be sure to do your own research and make sure it has a proven track record for being effective against bed bugs. Some things to look for:

- The material the encasement is made of must not allow bed bugs to feed through it.
- The encasement must use an appropriate zipper and have a zipper locking mechanism.
- Finally, the material must be durable enough to withstand normal wear and tear without ripping.

Bed Bug Interceptors

Bed bug interceptors are simple devices used for isolating furniture and for bed bug monitoring. You will need a set of bed bug interceptors if you are isolating your bed. I also recommend using them to isolate other furniture after it has been treated. These are fairly inexpensive, at a cost of about $12-$19 for a set of four. However, if you need a lot of them the costs can add up quickly. A typical two-bedroom apartment or home will typically require at least twelve bed bug interceptors. Fortunately, you can make your own bed bug interceptors at home from easily obtained materials. Keep reading for more about that.

Want to Purchase Interceptors?

A company named Hot Shot manufactures a bed bug interceptor that is sold in Home Depot. I have never used it before and can't vouch for it but these are fairly simple contraptions so I wouldn't hesitate to grab them if you want to go that route. Lowes may also carry this product but the one in my area did not have it in stock.

You also have the option of purchasing a more reputable bed bug interceptor with a lower price from Amazon.com. Three that I have experience using are:

- Climbup® Insect Interceptor Bed Bug Trap

- BEAP 10013 4-Pack Detector Bed Bug Coaster Trap
- Aspectek Bed Bug Trap, Interceptor and Insect Monitor Control

Want to Make Your Own Bed Bug Interceptor?

Here is a great video provided by the IFA Extension at the University of Florida that shows just how easy it is to make your own bed bug interceptors.

https://www.youtube.com/watch?v=Jjc4CD4U4uQ

Vacuum Cleaner

Some people I talk with are nervous about contaminating their regular household vacuum cleaner and prefer to purchase a unit that is solely dedicated to eliminating bed bugs. While you don't have to buy a vacuum cleaner that you will use solely for bed bug removal, you might choose to, simply because some types of vacuum cleaners are more difficult to clean than others. If your regular vacuum is designed in such a way that it's not easy to contain its contents, using it for bed bug removal could increase the likelihood of spreading an infestation rather than containing it.

If you decide to purchase a "bed bug dedicated" vacuum, I recommend you get a shop vacuum with 4.5 horsepower and a HEPA filter. These units have adequate suction power, are easy to clean, and cost around $75 dollars or less. Here are examples of some good ones:

- Amazon: Shop-Vac® 5-Gallon 4.5 Peak HP Stainless Steel Wet Dry Vacuum
- Home Depot: RIDGID 9-gallon Wet/Dry Vacuum
- Lowes: Shop-Vac 10-Gallon 4.0 Peak HP Shop Vacuum

Brilliant Frugal Alternative

I started using an unconventional method a few years ago that I read about on MyPMP.net. This method involves using a knee-high nylon stocking as a primary filter to catch bed bugs and their eggs in the vacuum hose before they have a chance to reach the inside of the vacuum cleaner. Not only does this technique provide you with a means of careful disposal, it also allows you to inspect the contents you just vacuumed up. I haven't been able to find the article again since reading it, but this method has been given the thumbs up by leading bed bug experts.

The instructions are simple. Just stuff the knee-high stocking, toe end first, down the hose of the vacuum. Wrap the outer edge of the stocking around the top of the vacuum nozzle and attach the crevice tool over the top of it to hold it in place. The stocking will catch bugs, debris, and anything else that enters. When you are finished vacuuming, the stocking can be removed, inspected, and thrown away inside a sealed contractor bag. Just don't turn the vacuum off until you have safely removed the stocking, or the bed bugs could fall back out.

This method will allow you to safely use your household vacuum without fear of contaminating it with bed bugs.

Steam Cleaner

A steam cleaner is a wonderful tool for instantly killing bed bugs that you are not able to reach with a vacuum cleaner. The downside is that they are pretty expensive if you plan to buy one. Fortunately, I have a few alternatives for you.

Key things to look for in a steamer if you are buying a steamer:

- Dry (or "dry vapor") steam (This is important.)
- Heat at the tip is >200F degrees per manufacturer
- Large water chamber to cut down on starts and stops in the process
- A good warranty on the boiler

There are many steamers on the market but not all of them are useful for killing bed bugs. This is especially true when it comes to lower-end consumer units in the $50-$250 price range. The steamer I use is the Vapamore MR-100, as it is the most affordable dry vapor steamer that meets all of the criteria on our list.

You can see the current price and purchase the Vapamore MR-100 from Amazon. By shopping around, you may find products that work as well or better. If you do, please let me know.

Frugal Alternative 1:

Before ASG dust, I would have said that a steamer is necessary for any infestation regardless of the circumstances. Diatomaceous earth just wasn't effective enough to rely on for spot treatments. With ASG dust in the picture, however, not every infestation will require a steam cleaner. You'll need to make the determination yourself by examining your infestation and deciding if there any areas that will require steam.

Frugal Alternative 2:

You may be able to rent a steamer from a local tool and equipment rental shop. The closest rental shop to me that offered these units was about 40 minutes away, so I opted to buy my steamer instead. You may be able to find a rental shop closer to your home. I have been told that a steamer rents for as little as $50, depending on the length of time it is needed.

Dissolvable Laundry Bags

Laundry is a huge part of treating a bed bug infestation, and it involves moving a lot of potentially infested items through your house to the laundry room. At the very least, everything should be placed in contractor bags and then carefully emptied into the washing machine. The empty contractor bag then needs to be disposed of properly.

You can achieve laundering your items safely if you are careful. However, if you would like to minimize the chances of slipping up, you can use dissolvable laundry bags.

Dissolvable laundry bags:

- Are biodegradable. They have been used in the healthcare industry for decades.
- Dissolve completely in water at temperatures of 140 degrees F or higher.
- Will not damage clothes or equipment and leave no residue behind.

I have only used these bags once, and they did make things easier. More importantly, they will provide you with peace of mind during a difficult time. I believe they are worth every penny if you can spare the extra $60. However, these are an optional item and you don't need to buy them if you don't want to.

If you decide to use them, you'll need to add them to your Amazon order because they won't be available locally.

Nuvan Prostrips

Like most other pest control chemicals, Nuvan Prostrips work by paralyzing an insect's nervous system, which causes it to die. If used properly, these strips are not dangerous to humans or domestic animals.

I have to admit that I have never used Nuvan Prostrips or any similar products because I like to treat infestations naturally, but I cannot deny how useful they are in decontaminating belongings that cannot be treated by any other means. You'll see on Amazon.com that they have great reviews. So if you have items that need to be treated that can't be steamed, washed, incubated for 18 months, or sprinkled with ASG dust, then Nuvan Prostrips would be your best option. All you need to do is place a Nuvan Prostrip into a contractor bag with the items you need to treat. Make an air-tight seal, give the Prostrip a few days to work, and the product will take care of the rest.

As of the time of this writing, Nuvan Prostrips cannot be shipped to California, Connecticut, or New York due to regulations.

Portable Bed Bug Heaters

Portable heaters are available to treat smaller items in your home. This is a more expensive option but bed bug heaters are extremely effective. They come in various sizes and are an efficient way to kill all bed bugs in all life stages, including eggs. They are useful for harder-to-treat items such as books, files, electronics, and other household items. They work because temperatures between 130 and 150 degrees kill bed bugs without damaging your belongings. These heaters can also be utilized to treat luggage when you return from traveling to ensure bed bugs are not transferred into your home. The downside is that they are quite expensive.

I never needed to use a bed bug heater but here are a few products I have heard good things about:

- Zap bug heater
- Zap bug oven 2
- Thermal strike ranger
- Thermal strike expedition
- Zap Bug Room

Bed Bug Traps

The main purpose of a bed bug trap is to monitor for the presence of bed bugs, not to control an infestation. That's the reason it is virtually impossible to find a bed bug trap with good ratings. Do a quick search on Amazon to see what I mean. Unsuspecting consumers buy bed bug traps with the expectation that the trap will attract and kill mass quantities of bed bugs, but that is not what they are meant to do.

However, if your purpose is to monitor for bed bugs, there are various devices on the market that will do that for you. Many of them attract bedbugs by emitting carbon dioxide (CO_2).

Researchers from Rutgers University recently designed a bed bug trap that utilizes dry ice, which releases CO_2. I use their design to make my own traps. In my experience, these homemade traps work better than store-bought ones, because with a homemade trap you can control the amount of CO_2 that is emitted, and you can replenish the CO_2 source when it is depleted. That way, the traps remain effective longer.

Bed bug traps consist of a simple container that allows bed bugs to climb inside where they become trapped. The containers are similar to bed bug interceptors, which we discussed previously. The only addition to that concept is the lure or bait, which is CO_2 on the traps I'm about to describe for you. You could modify the technique below by using any leftover interceptors.

Making a Dry Ice Trap

1. The first item you will need is container that will allow bed bugs in, but not allow them to crawl back out. Researchers from Rutgers University recommend using a 64-ounce double-bowl cat feeder. I have found that a round single-bowl cat feeder works just as well. You can also use leftover

bed bug interceptors or even homemade versions of the interceptor if you made those.

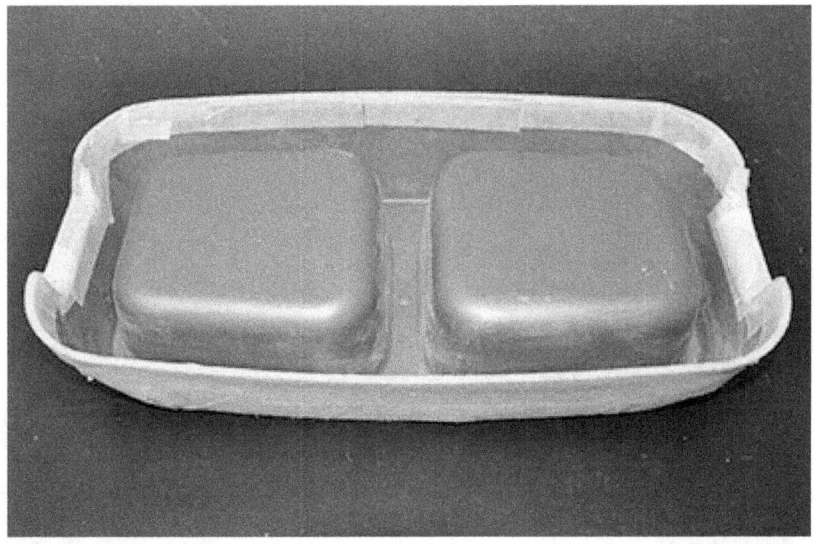

Photo credit: Rutgers University NJ Agricultural Experiment Station
https://njaes.rutgers.edu/pubs/fs1117/

2. Purchase a 1/3-gallon insulated jug. You can find these at any local retail store for about five dollars. This is the container that will hold the dry ice. It has a perfect design so that it will release carbon dioxide at about the same rate as would two people sleeping in a bed.
3. Purchase a small bottle of talcum powder. You may already have this but it can also be purchased at any local retail store. Talc is most commonly available as baby powder products in retail stores.
4. Bed bugs cannot climb smooth surfaces so you'll need to attach a piece of white fabric to the exterior surface with masking tape. This will allow easy entry to any bugs searching for a meal. Be sure to use masking tape, not smooth Scotch tape or packing tape, because bed bugs cannot

climb those clear, smooth surfaces. Be sure to attach the upper and lower edges of the fabric tightly so bed bugs won't be able to hide beneath the fabric.

5. I have never tried this, but researchers at the agricultural experiment station at Rutgers say that you can also roughen the outer surface with sandpaper to create a surface that bed bugs can climb.

6. With a cotton ball, apply an extremely thin layer of talc to all inside surfaces of the trap. This will give extra assurance that the bed bugs will not be able to climb back out of the trap once they get inside it.

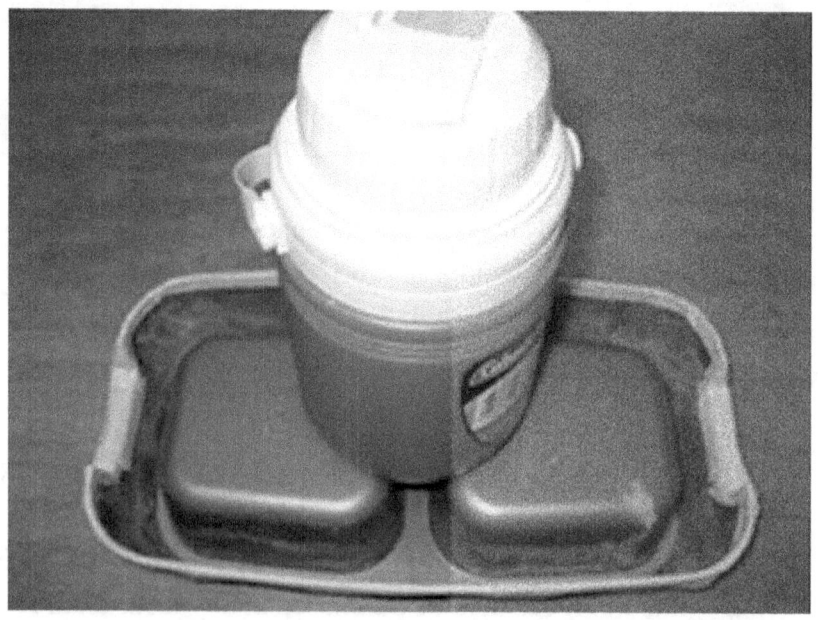

Photo credit: Rutgers University NJ Agricultural Experiment Station
https://njaes.rutgers.edu/pubs/fs1117/

Acquire Dry Ice

1. Dry ice is a fairly common product so it should not be difficult to find. Just search the internet for "dry ice" plus

your Zip Code or the name of your town and you should get multiple hits within a short distance of your neighborhood.
2. Dry ice is inexpensive, usually costing $0.99 per pound. A good amount to start with would be about ten pounds per room.
3. Unused dry ice can be placed in a foam cooler and stored in a freezer temporarily for up to three or four days. That is why you want to keep your supply limited to about ten pounds per room. (You'll be using 2.5 pounds per night, per trap).
4. **Always use caution** and follow the manufacturer's instructions when handling dry ice. Dry ice holds a temperature of almost -110 degrees Fahrenheit, and it can therefore be very dangerous if you are not careful.

Finish Making the Trap

1. You'll be able to fit about 2.5 pounds of dry ice in a 1/3-gallon jug. Fill the jug up and leave the nozzle slightly open. Position the jug in place over the upturned cat bowl feeder as shown above.
2. Position the trap near your bed and/or other areas where the most bed bug activity seems to occur. Since the trap has a 10-12 hour lifespan with 2.5 pounds of dry ice inside it, you should wait until the early evening before filling the jug and placing the trap. Bed bugs prefer to feed at night so this is the time the trap will be most effective.
3. Repeat these steps to construct traps for monitoring in additional rooms, with a limit of one trap per room.
4. Do not use ceiling fans during the time the traps are active.

The Next Morning

1. Examine the trap when you wake up in the morning. The folks at Rutgers University did a fantastic job with the design

of this trap so you can expect to see much better results than with store-bought traps. People have reported anywhere from three to 150 bed bugs being trapped per night. Much of that will depend on the infestation levels.
2. Handle the trap with care because there still may be bed bugs on the outside that have not managed to trap themselves inside yet.
3. Dispose of the bed bugs in the trap by drowning them in the toilet. Again, just be careful there are not any hitchhikers on the outside of the trap before you move it from one room to another.
4. If there are no signs of bed bugs in the trap don't be discouraged. This is an effective trap, so no bed bugs is a good sign.
5. Continue to refill the trap with dry ice each evening and monitor the room for the next few days, even if you found no signs of bed bugs at first.
6. Remember that a bed bug trap is helpful in monitoring an infestation but it cannot be relied on to determine with complete certainty that an infestation is eliminated. You'll need to continue monitoring and inspecting daily before you can let your guard down. The biggest concern here is with eggs that haven't hatched yet. All signs could point to the infestation being wiped out, but a few newly hatched eggs could change things quickly if you don't maintain the treatment procedures we talked about in the guide that accompanies this manual.
7. When you are finished with the trap, use the procedures you learned in my guide to make sure it is decontaminated before removing it from the infested room. If you don't want to go that route, you can simply dispose of it in a sealed contractor bag.

Other Bed Bug Products You May Have Heard About

There are many common recommendations you will find online when it comes to getting rid of bed bugs. Below is a list of things you may have heard about that are not used in the treatment plan I use. Some of these items work fine, just not as well as the alternatives used in my method. Other items may have been excluded because they are dangerous, too expensive, or just simply don't work.

- **Diatomaceous earth (DE):** I used DE to help treat the bed bug infestation in my own home. However, with ASG dust in the picture, there just isn't any reason to use it anymore, as DE is not nearly as versatile or effective as ASG dust. While it does have its place in bed bug control, 100% amorphous silica desiccant dust (CimeXa) is a far superior solution.
- **Isopropyl "rubbing" alcohol:** Alcohol is a bed bug contact killer, meaning it needs to be applied directly to the bugs in order to be fatal. It is cheaper than most other insecticides and safer in many ways. However, alcohol is extremely flammable. There have been cases in the news recently of people burning their houses down while treating for bed bugs.
- **Bed bug-detecting canines**: Yes, you read that correctly. Dogs are now being trained to detect bed bugs. I am sure those bloodhounds and beagles are wonderful at what they do, but it is overkill if you ask me. The inspection and detection methods in this book are tried, true, and tested. Besides, hiring a company to bring bed bug-detecting dogs into your home will cost anywhere from $200-$1,000 dollars.
- **Heat Treatments**: You can hire a company that specializes in killing bed bugs with a heat treatment system. You house will be sealed and baked to 120 degrees for a couple of hours. If you decide to hire a company to assist with your bed

bug infestation, this is the direction I would steer you. While it is one of the more effective methods, it is also one of the most expensive. Prices for home heat treatments to kill bed bugs can cost several thousands of dollars.

Unfortunately, cranking up your furnace or using portable heaters to create the same effect will not work. Aside from being ineffective, that would also be very dangerous due to toxic fumes and risk of fire. This is truly a treatment procedure that should be left to the professionals.

Items to Avoid at All Costs

Store-bought chemicals: Despite many companies trying to cash in on bed bug removal, studies have shown conclusively that store-bought chemicals DO NOT WORK. They will make your bed bug infestation worse and more difficult to treat.

Bedbug fogger: Foggers do not work. Bed bugs are highly resistant to the chemicals they use. A fogger will just drive the bugs deeper into the cracks, crevices, and walls of your home.

Freezing: Freezing bed bugs is possible but they are much more difficult to freeze to death than they are to heat to death. You will need to maintain a temperature of zero degrees Fahrenheit for a minimum of four days, and this is simply not possible to do in your home. Likewise, using your climate to freeze bed bugs outside is tricky because of sunlight, humidity, and temperature variations. It is unlikely that outside temperatures will stay cold enough for the period of time required to kill all bed bugs and their eggs.

Bed Bug Supply Summary and Checklist

1. Online purchases (with one- or two-day shipping preferred so you can get started more quickly)
   - Two or three 4-ounce bottles of CimeXa ASG dust from Amazon.com
   - Hand duster from Amazon.com
   - Mattress and box spring encasements
   - Dissolvable laundry bags (optional)
   - Nuvan Pro Strips (optional)
2. Bed bug interceptors
   - Add to Amazon.com, Home Depot or Lowes order, or make your own according to the instructions in this manual
3. Make a decision on vacuum cleaner
   - Use your own?
   - Buy a dedicated bed bug vacuum?
   - Knee-high stocking filter technique?
4. Decide if you will need a steam cleaner for contact killing
   - Vapamore MR-100 from Amazon.com
   - Rent a similar one from a local tool/equipment rental facility
5. Take inventory and purchase any items on the list below that you don't already have.
   - Shop online at HomeDepot.com or Lowes.com and let them pick the order for you to save time

Bed Bug Treatment Supplies	
Contractor bags and/or XL, XXL Ziploc Bags	Utility knife
Sponge	Rags/paper towels
Dish detergent	3-4 rolls of Scotch or packing tape
Lysol and Murphy's Oil Soap	White sheets, pillow cases, and cotton blankets
Paintable latex caulk and caulk gun	Duct tape
Joint compound	Bucket
Wood filler (if needed)	Mineral oil (if you are isolating your bed)
Expanding foam (if needed)	Vaseline (if you are isolating your bed)

6. Purchase the items you will need for a bed bug trap if you will be using one for post-treatment monitoring. This can be done at a later date. I would put all my energy into the first treatment before worrying about a bed bug trap.

www.ingramcontent.com/pod-product-compliance
Lightning Source LLC
Chambersburg PA
CBHW061441180526
45170CB00004B/1513